CW01379259

Insight in Innovation

Insight in Innovation

Managing innovation by understanding the Laws of Innovation

Jan Verloop

With contributions from Hans Wissema
Delft University of Technology

Shell Global Solutions

2004

ELSEVIER

Amsterdam • Boston • Heidelberg • London • New York • Oxford
Paris • San Diego • San Francisco • Singapore • Sydney • Tokyo

ELSEVIER B.V.	ELSEVIER Inc.	ELSEVIER Ltd	ELSEVIER Ltd
Sara Burgerhartstraat 25	525 B Street, Suite 1900	The Boulevard,	84 Theobalds Road
P.O. Box 211,	San Diego,	Langford Lane	London WC1X 8RR
1000 AE Amsterdam	CA 92101-4495	Kidlington,	UK
The Netherlands	USA	Oxford OX5 1GB, UK	

© 2004 Elsevier B.V. All rights reserved.

This work is protected under copyright by Elsevier B.V., and the following terms and conditions apply to its use:

Photocopying
Single photocopies of single chapters may be made for personal use as allowed by national copyright laws. Permission of the Publisher and payment of a fee is required for all other photocopying, including multiple or systematic copying, copying for advertising or promotional purposes, resale, and all forms of document delivery. Special rates are available for educational institutions that wish to make photocopies for non-profit educational classroom use.

Permissions may be sought directly from Elsevier's Rights Department in Oxford, UK: phone (+44) 1865 843830, fax (+44) 1865 853333, e-mail: permissions@elsevier.com. Requests may also be completed on-line via the Elsevier homepage (http://www.elsevier.com/locate/permissions).

In the USA, users may clear permissions and make payments through the Copyright Clearance Center, Inc., 222 Rosewood Drive, Danvers, MA 01923, USA; phone: (+1) (978) 7508400, fax: (+1) (978) 7504744, and in the UK through the Copyright Licensing Agency Rapid Clearance Service (CLARCS), 90 Tottenham Court Road, London W1P 0LP, UK; phone: (+44) 20 7631 5555; fax: (+44) 20 7631 5500. Other countries may have a local reprographic rights agency for payments.

Derivative Works
Tables of contents may be reproduced for internal circulation, but permission of the Publisher is required for external resale or distribution of such material. Permission of the Publisher is required for all other derivative works, including compilations and translations.

Electronic Storage or Usage
Permission of the Publisher is required to store or use electronically any material contained in this work, including any chapter or part of a chapter.

Except as outlined above, no part of this work may be reproduced, stored in a retrieval system or transmitted in any form or by any means, electronic, mechanical, photocopying, recording or otherwise, without prior written permission of the Publisher.
Address permissions requests to: Elsevier's Rights Department, at the fax and e-mail addresses noted above.

Notice
No responsibility is assumed by the Publisher for any injury and/or damage to persons or property as a matter of products liability, negligence or otherwise, or from any use or operation of any methods, products, instructions or ideas contained in the material herein. Because of rapid advances in the medical sciences, in particular, independent verification of diagnoses and drug dosages should be made.

First edition 2004

Library of Congress Cataloging in Publication Data
A catalog record is available from the Library of Congress.

British Library Cataloguing in Publication Data
Verloop, Jan: Insight in innovation: managing innovation by understanding the laws of innovation
1. Creative ability in business   2. Industrial management
I. Title   II. Wissema, Hans   658.4'063

ISBN 0-444-51683-2

∞ The paper used in this publication meets the requirements of ANSI/NISO Z39.48-1992 (Permanence of Paper).
Printed in The Netherlands.

Contents

About the book ix
Foreword xi
Acknowledgements xiii
Introduction xv

Chapter 1 The Innovation Process

Lessons from the first 500,000 years of innovation 1
Fire 1
Columbus 2
Edison 4
The universal stages of innovation 5
The three stages 7
The process of innovation 11
The classic innovation model 11
The bridge-building innovation model 15
The innovation processes through history 17
Summary 20

Chapter 2 The Innovation Spectrum

Classes of innovation 21
Two types of innovation 22
Outside-the-box innovation 25
Ways of innovation 29
Top-down and bottom-up 29
Alliances or all-alone 31
Innovation zones 33
Innovation and transformation 35
Internal innovation 35
The dynamics of innovation and transformation 41
Summary 43

v

Chapter 3 Managing Innovation

Innovation is a business process 45
The innovation supply chain 46
The need for staging.. 47
Stages and tollgates 49
The stages 49
The tollgates 54
The roles of the innovation manager 58
Managing the innovation supply chain.................... 59
Executing the innovation strategy 61
Managing the innovation portfolio......................... 63
Innovation infrastructure 65
Understanding innovation 65
Organisational principles 66
Reward and recognition 68
Innovation culture 68
Summary 70

Chapter 4 Innovation and Entrepreneurship

Implementing the innovation options 73
The 'Valley of Death' 73
Monetising the options 74
The final hurdles... 82
Timing 82
Infrastructures.. 85
The fight back of the old technologies 86
Emotions... 87
Summary 90

Chapter 5 The Value of Innovation

Innovation creates options 91
The three value domains 92
Assessing the value of innovation............................ 98
The value of an option... 99
The value of a portfolio of options 102
Innovation creates intellectual capital 106
A segmentation model for intellectual capital 107
The value of intellectual capital 110
The value of intellectual property........................ 112
Summary 116

Contents

Chapter 6 Sustainable Innovation

Sustainability and innovation 117
The challenges for sustainable innovation 118
The sustainable innovation model 124
Learning from the future 127
Trends 128
Techniques 130
Assessing sustainable innovation 131
Criteria for sustainable innovation 132
Sustainable innovation creates societal capital 134
Summary 136

Chapter 7 Innovation and the CEO

The final law 137
The Laws of Innovation 139

References 143
Index 147

About the book

Innovation has always been an important topic for Shell Global Solutions. Indeed, the creation of Shell Global Solutions was a significant innovation in its own right, and the innovations that were built into it are still the cornerstones underpinning our commercial success and the value propositions we make to our customers.

Innovation is an ancient art, but managing innovation is a relatively young management technique and has received much less attention than other aspects of innovation such as creativity, entrepreneurship or venturing. Managing innovation in such a way that it becomes an effective business process to achieve strategic objectives set by company management is still in its early stages of development. This book provides insight into the management process for innovation and the role of innovation in creating intellectual capital and supporting sustainable development. These 'insights in innovation' are based on many years of industrial experience accumulated by the author and his many colleagues who have supported his efforts, and these have been further enriched through intensive contacts with the academic world.

Personally I'm very pleased that my predecessor agreed to sponsor this book. It brings me back to 1998 when two founding documents for Shell Global Solutions had to be written. I drafted the initial business plan for the fledgling company and Jan Verloop drafted the Technology Vision, our technology and innovation strategy document. Some of the concepts and pictures in that document reappear in this book; the ideas and concepts remain as challenging and valuable as ever.

Greg Lewin
President Shell Global Solutions

Foreword

A sustainable planet needs sustainable innovation

Innovation is not, in my view, a 'Eureka' moment or an 'aha experience', but the application of new thinking to how we work, what we make, the service we provide or how we enjoy life.

An innovative country is one that has found the key to where science meets business (a one-liner of the Society of Chemical Engineers in The Netherlands) and new things happen because there is a good investment – or 'refurbishing' – climate.

An innovative company can apply new thinking faster than its competitors or surprise its customers with 'aha experiences'. This means having the right people, principles, managerial processes and determination. Having all these elements helps a lot – and what helps even more is having senior people who drive them forward with personal passion and courage.

Innovation has always been a key business issue for Shell; indeed, it was one of the founding elements of Shell, as exemplified by the story of 'Murex and the red tins' in this book. What we at Shell are trying to do now is to make innovation an integral part of the way we do business. There were times when research was very much a world on its own, a splendid and creative world to be sure, but even the best ideas are no good if business does not use them and they never reach the customers who can benefit from them.

Leaders should give focus. By focus I mean not just 'what' but also 'how'. For instance, how sustainable development can be put into practice to stimulate or check innovation from various perspectives: economic, social, environmental, medical and so on.

Jan Verloop's book is practical and inspirational. It offers the historical context in which innovation is placed. The stories about Columbus and Edison contain essential lessons for the innovation manager of

Insight in Innovation

today, and also show that the way innovation works changes over time; each era in history needs to find its own way of solving problems. This is an important message. Managing innovation better is not only a matter of continuously improving the innovation process, it is also about re-inventing innovation when times are changing.

A good example of this is the link between innovation and sustainable development. Creating a more sustainable world is one of the great challenges of our time, and the hydrocarbon industry has a major role to play. Sustainable development is embedded in Shell's business values, and we at Shell are keenly aware that this means we shall have to be very innovative to make a real contribution. A more sustainable world will need significant changes. Technology can help – for instance, how we use technology in business and the choice of products and services we offer to our customers. Innovation is the key to creating change, and only innovative companies can be successful in a changing world. But being innovative will not be sufficient by itself: the message of this book is that the way we innovate also needs to be sustainable; a sustainable planet needs sustainable innovation.

Of course, I particularly enjoyed the examples in which Shell played a role. Shell Global Solutions, Shell Hydrogen and GameChanger are all examples of how we have responded in an innovative way to the challenge of doing business in an environment that is continuously changing. Even if you think you already know a lot about innovation, this book will help to innovate your innovation.

Jan Verloop has put his passion and determination into this book. I will conclude with one more thing I liked about it. Management books can be boring when they only contain descriptions and instructions on how to make business processes more efficient and effective. This book includes emotions as an integral factor in the business process of innovation and convinces the reader that without a vision, without passion and without determination, innovation will not be successful. As you read it, I hope that your thoughts about innovation will undergo this innovative process.

Jeroen van der Veer
Chairman of the Royal Dutch/Shell Group

Acknowledgements

Like most books, this book needed support from many sides, and, appropriately for a book on innovation, the support came in three phases: the idea, the development and the investment phase.

In the idea phase I'm indebted to Professor Wissema. Hans reads innovation and technology management at Delft University of Technology and in that context he was interested in the Shell experience. It was he who suggested to write a book on the topic. The initial idea was to jointly write two books in one cover to bring the academic and the industrial worlds together in one book, but the contents decided otherwise and there will now be two separate books. Hans is still writing his book, but his contribution during our many discussions and his challenges on the drafts have been very valuable to me.

This book is based on my involvement with innovation in Shell, and I would like to thank all my colleagues who contributed in so many ways to my understanding of innovation. In particular, I want to mention the Innovation Coalition, the loosely structured collection of committed GameChangers in Shell that continuously tries to improve the ways of innovation. Several of them have contributed to the development phase of the book by reading sections or listening to and commenting on my train of thought. Also the interviews with Jeroen van der Veer, Greg Lewin and Adri Postema provided valuable contributions to highlight key experiences that found their place in the box stories.

For the investment phase, I would like to thank the presidents of Shell Global Solutions. Hans van Luijk, the first president, introduced me to Professor Wissema and it was at his request that in 1998 I drafted the Technology Vision, the founding innovation and technology strategy document for Shell Global Solutions. Hans, together with his successor Rob Routs, signed off the document and many of the ideas in that document are now reflected in this book.

I'm also very much indebted to Michiel Boersma, the third president, who agreed that I should write this book, and my final thanks go to Greg Lewin, the current president, who has continued to sponsor the book's production.

A special word of thanks must go to Gill Rosson, who skillfully and supportively edited my manuscripts and you, my dear reader, may thank her for that as well.

Introduction

In the mid-nineties, the Royal Dutch/Shell Group of companies went through an in-depth organisational change based on a fundamental reorientation of the way it conducted its business. Subsequently many of its business units had to reinvent themselves and redesign their business process to fit and thrive in the new environment. This was a major innovation effort in its own right, but the business units also had to rethink their innovation efforts on the products and services they delivered to their customers. This book is based on the experience of this transformation in innovation, but also on the shared experiences with the rest of industry and is embedded in an extensive body of academic research.

This book is primarily about managing radical innovation in large companies. In large companies radical innovation almost unavoidably leads to erosion of established positions and thus to internal resistance. There are many hurdles to radical innovation inherent to being big and there can easily be a disparity between intention and action. Consequently, creating and implementing a coherent innovation strategy in a large company that has a genuine impact on the bottom line is not simple, but requires sound business processes and dedicated attention from top management.

Innovation is a complex process and typically the road from idea to market is tortuous and full of unexpected surprises. For hundreds of years the innovation process had a large degree of randomness and this was thought to be part of the natural flow of the innovation process. But innovation can and needs to be managed. Innovation is not a magic ingredient that some companies have and others don't. It is a business process like credit management, strategy development or competitor intelligence that can be neglected or managed to its potential. However, probably because of its complexity, innovation is one of the last business processes still not subject to systematic management everywhere. Since the industrial revolution, manufacturing processes, investment decisions, project implementation etc. have been the subject of many management theories and have well-established best practices, but the history of innovation management only covers about 20 years. Drucker called the management of business processes one of the great innovations of the twentieth century, but innovation is one of the last areas where it is applied. In order to be innovative as a company, it is not good enough to be creative, or research intensive or entrepreneurial. If the whole process from generating novel ideas to implementing in the market place is not managed properly and is not aligned with the company strategy, the innovation effort will not be effective. It may create centres of excellence in the company and provide good PR stories, but the value of the company will not improve.

The starting point for good innovation management lies in understanding the nature and basic principles of the innovation process and this book gives the essential characteristics of the innovation process that an experienced manager should understand in order to master his role in the innovation process, either as a CEO, a research, marketing, innovation manager or otherwise. On the basis of these fundamentals the specific business processes and governance structures can be designed and these will differ from company to company, reflecting their overall management style and company culture.

We have called the few elements that are basic and intrinsic to good innovation management the 'Laws of Innovation', and believe that if a CEO understands these laws and makes sure they are built into the company innovation process, he can be confident that the governance of the innovation effort is sound and effective.

This book is not about providing a series of recipes on innovation management or a collection of case stories on how to do innovation or not. The few examples given are well known innovations from (Shell) history and all of them have been described before in the literature. However, in this book we do not focus on the brilliant result or failure of the innovations, but on the process of innovation in order to understand the features of a well-managed innovation effort.

This book has been written around six themes, roughly coinciding with the first six chapters. The first three themes are about understanding the innovation process and how to manage it; the last three themes are on specific topics that link innovation respectively to risk and entrepreneurship, value creation and intellectual capital, and sustainable development.

Introduction

The industrial innovation process

The first theme is on understanding the nature of the industrial innovation process and how it has developed through history. Over time the innovation process has changed fundamentally. Before the industrial revolution, innovation existed, but it missed the systematic application of science and technology. The industrial era has brought the classic approach to innovation that started with scientific discoveries. But since then the innovation process has changed from a technology- to an opportunity-driven process, aimed at finding and creating business opportunities based on new technology/market combinations. Also the development process from idea to market has grown in complexity and needs involvement from an increasing number of stakeholders.

A critical step in managing innovation is to understand that the path from idea to market needs to be staged. The number of stages is optional, but the three fundamental stages are generating the idea, demonstrating the concept, and investing to extract the benefits from the market. The three stages in the innovation process are universal and can be recognised throughout history, for instance in the discoveries by Columbus and the innovations by Edison. Good innovation management recognises these stages and manages them differently in line with their specific requirements. But equally important is to manage the whole process in its totality and avoid the stages becoming independent and operating as separate silos. Innovation management is managing the innovation supply chain.

Innovation is a business process to create change

The basic reason for innovation is the need to change. Usually this change equates to product or business improvement, but sometimes the change needs to be more fundamental. The need for radical change can be for internal reasons based on strategic objectives to step-change the competitive position or for external reasons because the business environment has changed so much that the company has to reinvent itself. One of the main benefits of radical innovation is that it makes a company more aware of and resilient against external changes. It is often said that innovation is the key advantage in the 'survival of the fittest', but be aware that fittest does not simply mean strongest. The original words of Darwin describe the purpose of innovation very well:

> *"It's not the strongest who survive, but those most responsive to change."*

From a management perspective only two types of innovation are relevant: 'outside-the-box' and 'inside-the-box', depending on the degree of change required and the business process to be adopted.

Inside-the-box innovation is incremental innovation in products, processes and services to support and develop existing businesses. It operates at product strategy level, is fully executed within a business unit, and this type of innovation is a 'must do' activity, required by any business.

Outside-the-box innovation is game-changing innovation to create change for strategic reasons or in response to changes in the business environment. It is a corporate activity and operates at company strategy level. Game-changing innovation is not a necessity, but a strategic choice and the CEO needs to decide whether and how to use this tool to achieve the required changes.

Managing the innovation process

Innovation needs both creativity and investments. Creativity needs openness, freshness and a willingness to say yes; it thrives on a bit of chaos. Investments need discipline, sound analyses and a willingness to say no. The element of chaos gives innovation a bad reputation amongst many CEOs who equate chaos with the absence of control and good management. Separating the different stages and keeping creativity, analyses, discipline, etc. where they belong is an essential part of innovation management.

Managing the innovation supply chain in line with the specific characteristics of each stage and maintaining the cohesion and momentum in the whole chain is the first challenge of the innovation manager. The second task for the innovation manager is to ensure that the tollgates between the stages are properly controlled. At each stage the innovative idea has to meet certain specific criteria and deliverables and these have to be assessed by the custodians of the tollgate. For game-changing innovation managing the tollgates is part of corporate management. The third role of the innovation manager is to optimise the value of the innovation portfolio in relation to the innovation

Introduction

effort and available capabilities, and ensure that the portfolio has the potential to meet the company objectives.

Innovation and entrepreneurship

Crossing the last stage of the innovation process is often the hardest. The first stages of the innovation process have demonstrated that the innovation has the potential to become a valuable business opportunity. It has created an option and the challenge is to find a 'buyer' for that option. The best way is to 'sell' it inside the company to a business with a related portfolio of activities. However, most business managers will react reluctantly to a game-changing proposition, because the risks are high and the profits come in slowly. As a result many innovations do not pass this tollgate, stay in limbo and gradually decay in the 'valley of death'.

If an entrepreneur cannot be found inside the company, an alternative option is to find one outside the company. This approach may reduce the value of the option for the company significantly. Part of the value will get lost in order to create a win-win situation for the new partner in business unless the partnership adds value to the option.

The value of innovation

At the current state of the art, assessing the impact of innovation on the top- or the bottom-line of a company is one of the weaker parts in the management process. However, you get what you measure. The best way to measure the value of game-changing innovation is to estimate its option value. The option value reflects the potential business value of an idea, discounted for the chances of success to make it to the launch and the uncertainties in the development path and market conditions. The option value allows players to track the progress in adding value to the idea and facilitates stopping a project on rational grounds without attaching blame to the players.

The innovation capability is one of the intangible assets of a company. Not many companies quantify the value of their intangible assets, which is surprising because for many companies the intangible part is more important then the tangible part. However, if all intangible assets were valued, the platform for managing innovation would be more robust. Innovation creates intellectual capital and, by valuing it as such, it can be compared to the other assets of the company and establish its proper position in the priorities of a company.

Sustainable innovation

The innovation process has changed in the post-industrial era. In the industrial age innovation started with scientific discoveries and science and technology were equated with progress. In today's world the attitude towards 'technological progress' is ambivalent and 'sustainable development' is emerging as the new core societal value. Thus innovation has to support sustainable development and this requires involvement from the customer and society at large. Ensuring societal acceptability is part of managing the innovation process.

To integrate sustainable development in the innovation process requires changes in the management model, the assessment and valuation of innovation, and a different interaction with the stakeholders. How this has to be done can only be described tentatively, because the whole concept of sustainable development is still in its early stage of development and the priorities of the various stakeholders differ.

It has been very tempting to equate 'sustainable innovation' to 'good innovation' and make it the first chapter rather than the last one. Although theoretically not without its merits, this approach would not have helped to make the special aspects of sustainable innovation clear. At this stage the insights into traditional innovation are more robust than those concerned with sustainable innovation, but understanding what sustainable innovation needs and how to execute it the right way will be the challenge for those companies that want to be *'most responsive to change in order to survive'*.

– 1 –
The Innovation Process

LESSONS FROM THE FIRST 500,000 YEARS OF INNOVATION

Innovation is an old activity, not as old as mankind, but could be 500,000 years or so old. In contrast, managing innovation is a relatively young discipline, effectively in existence for only tens of years. However, managing innovation is becoming more and more important, as innovation has to become more effective and the process of innovation becomes more complex. This complexity arises because of the increasing distance between the idea and the customer, and the increasing number of interfaces and players along the development path to the market. Complexity creates risks and managing innovation is about managing complexity and improving the odds of bringing an idea successfully to the market.

Innovation is often equated with creativity and inspiration, and from that perspective one may be inclined to leave innovation to the fortunate few who have got it. However, innovation is only 1% inspiration and 99% perspiration, and the perspiration part of the innovation process needs to be managed. Companies have to be innovative in order to survive and prosper, thus all companies should master the art of managing innovation.

We shall start with a short and simple overview of the history of innovation to learn from the past, identify the elements that have passed the test of time and those that have changed over time, as a basis for identifying and understanding the '**laws of innovation**'.

Fire

The control of fire some 300,000 to 500,000 years ago is one of the earliest, and possibly one of the most important, innovations of mankind. The ability to control fire is what uniquely distinguishes humans and animals. Indeed mastering fire may have initiated the rapid social and intellectual development of early mankind [1].

However, from a management perspective the innovation was relatively simple. The development path of the idea to the market probably was very short. The link between the invention on how to control fire and its use by a satisfied customer may have been almost immediate. The life of the innovator was simple. We can safely assume that he did not have to ask for a research budget, or write a customer value proposition on how to 'delight the customer'. The one thing that he most likely did have in common with all the innovators through the ages was that he had to take risks. He probably burnt his fingers a few times in the process, and the penalties for not controlling the fire would have been high. Taking risks is an essential part of the innovation process and it is a key role of today's innovation manager to identify, reduce and manage risks.

Columbus

Another famous innovation story from history, the discovery of America, provides a distinctive illustration of some of the essential features of the innovation process [2]. From an innovation management perspective, the discovery of America has room for improvement. Of course it is not fair to apply modern management techniques to one of the great historic discoveries, but there is some genuine learning in it.

For an innovation manager, this short story (see page 3) contains a few important learning points.

- Screening Columbus' proposal took the court of Castile more than four years. In a well-managed innovation process this should not take more than a few weeks or months.
- Columbus made a serious error in his calculations on the distance between Spain and The Indies, but by sheer coincidence still found The Indies at the expected spot on the globe. If the validating process had been more rigorous and Columbus had spotted his mistake from other evidence, King Ferdinand could have changed his business plan on what to do with the new territories.
- The king appointed Columbus as viceroy of the new territories, and, in doing so, made the classic mistake of appointing the discoverer as manager for the entrepreneurial phase of the innovation process; it took the king six years to correct this.

The Innovation Process

COLUMBUS AND THE DISCOVERY OF AMERICA

Columbus was born in 1451 in Genoa. After many trade voyages, he settled in Portugal in 1476 to study the works of Ptolemaeus and Toscanelli on the western route to The Indies. He developed his own theory on the distance to The Indies via the West and calculated that the western and eastern routes were about equal.

In 1484 Columbus made a proposal to King Johan II of Portugal to explore the western route. Subsequently, the same proposal was made to King Ferdinand of Spain in 1486. The proposal was initially rejected by the court, but finally accepted in 1490 when the king, counselled by Queen Isabelle, overruled the committee.

Columbus sailed West in 1492 and found The Indies at the expected position on the globe. Although he did not bring back gold or spices, Columbus got a hero's welcome on his return. He was appointed viceroy of the new territories and sailed again in 1494 with 17 ships and 1200 people to start colonisation. None of the new colonies was very successful and in 1500 Columbus was called back to Spain in disgrace and the new colonies were ignored for decades.

In 1502 Amerigo Vespucci sailed the coast of South America and realised that The Indies were a separate continent. His name was given to the continent by the cartographer Martin Waldseemüller on a new map in 1507. Years later the role of Columbus was rediscovered, but the name America was already widely and irreversibly adopted. However, even in the 17th century, Dutch maps carried the message that the name of the new continent should be Columbia.

- The whole innovation process took more than 30 years to go from idea to commercial success. There are many other examples in history showing that it takes 30 years or more to turn a major invention into a commercial success. In fact 30 is a fairly typical figure in industry for the number of years it takes for the innovation process to come to fruition. However, with the benefit of hindsight as well as insight into the innovation process, it would be quite possible to reduce the length of the innovation process for America from over 30 years to say five years.

The above 'points for improvement' are archetypal and it is the task of today's innovation manager to develop an efficient innovation process and achieve a time to market that is economically attractive. This requires good judgement on when to enter the race with respect to the required development time of the technology and the receptiveness of the market.

Edison

Edison provides another classic example of innovation. He is labelled in history as one of the great inventors, but it might be more appropriate

EDISON AND THE INNOVATION OF ELECTRIC ILLUMINATION

In 1802 Humphrey Davy discovered that an electric current passing through a metal wire could give light. In 1878 Edison joined the ranks of the many researchers trying to develop a light bulb with an acceptable operating life.
Edison wrote:

"I was more or less at leisure, because I had just finished working on the carbon button telephone, then this electric-light idea took possession of me. It was easy to see what the thing needed: it wanted to be sub-divided. The light was too bright and too big. What we wished for was little lights, and a distribution of them in people's houses in a manner similar to gas."

In less than one year he developed a bulb with a lifetime of 40 hours, obtained a patent and started the Edison Electric Light Company. He already had in mind that only if electricity were distributed in a way similar to gas, could the electric light bulb become economically attractive. So he improved the lifetime of the bulb to 300 hours, and at the same time developed a dynamo that generated electricity at 110 V, as well as a small distribution net, and demonstrated the whole system in 1880 with 425 lamps at Menlo Park.

Edison Electric Light Company financed another demonstration project, but the company did not believe that electricity would replace gas for illumination purposes and refused to make the investments in the factories to make light bulbs as well as generation and distribution equipment. Edison decided to finance the investments himself with his shares in Edison Electric. He became an entrepreneur and created three new companies to manufacture the equipment.

In 1892 Edison Electric changed its mind; the Edison Electric Illuminating Company was established and the new company obtained permission to light up Wall Street in New York.

The Innovation Process

to call him the greatest of innovators. He is famous as the inventor of the electric light bulb, but he should really be recorded as the innovator of electric illumination. His track record as an innovator is impeccable; he was both an inventor and an entrepreneur [3,4].

In only four years, Edison brought a significant invention, based on breakthrough technology and requiring a new infrastructure, successfully to market. Such a short time may well be a historic low for an innovation based on a disruptive technology. Simultaneously he also introduced a new standard: 110 V for electricity distribution in the USA.

Edison had a unique way of dividing a big problem into a series of small problems and solving them one by one, while keeping the whole value chain in mind. He approached the problem simultaneously from a technological and business perspective with the application in the market in mind.

The unique combination of qualities in one person – creative and analytical powers, entrepreneurship and the drive to succeed – are prime factors for this remarkable achievement. The above 'points of excellence' are archetypal and it is the task of today's innovation manager to ensure that both the creative and the entrepreneurial parts of the innovation process are linked and managed properly, and achieve with the innovation teams what Edison did on his own.

THE UNIVERSAL STAGES OF INNOVATION

Innovation does not mean the same thing to everybody. Many people confuse innovation with invention or think that both words stand for the same thing. Others equate innovation with creativity or entrepreneurship. Thus it is important that we first define what we mean by innovation.

The word 'innovation' stems from the Latin word 'innovare', which means 'making something new'. The first management definition of innovation is probably that given by the economist Joseph Schumpeter [5]:

> "The commercial or industrial application of something new – a new product, process or method of industrial production; a new market or source of supply; a new form of commercial, business or financial organization."

From this perspective the essence of innovation is the creation by entrepreneurs of 'Neue Kombinationen' (new combinations) between different technologies or technologies and markets.

Drucker [6] was the first to apply management techniques to innovation in a systematic way:

> *"Innovation is the specific instrument of entrepreneurship. It is the act that endows resources with a new capacity to create wealth. Innovation indeed creates a resource and endows it with economic value."*

Drucker also emphasises the role of the entrepreneur and specifically links innovation to adding value in the market place. Neither Schumpeter nor Drucker uses the word creativity in their definitions for innovation; the essence of innovation is the entrepreneurial activity that adds value to a new idea by bringing it to market.

In Shell a much-used definition is: 'Innovation is bringing an insightful idea successfully to the market'. Besides being short, this definition has several advantageous features; it brings across that:

- Innovation is a dynamic process; it is not an act, but a journey that needs to be travelled successfully to add value to an idea
- Innovation needs insight; creativity by itself is not enough, but the new idea needs to be grounded in a deep understanding of the technical merits and the requirements of the market place
- Innovation needs entrepreneurship to bring the idea to the market
- The market determines the fate of the innovation; successful innovation requires that the idea adds value to the customer.

Many more definitions can be found in the literature, but the key features of innovation are: insight, new combinations, entrepreneurship, adding value. This brings us to the first law of innovation:

Law I Innovation is the business process for creating new and insightful ideas and bringing them successfully to the market

The laws of innovation are business laws for management. The first law expresses innovation as a business process that covers both the creative as well as the entrepreneurial activities. In between these activities are the development activities that add value to the idea by creating the product or service that is needed or wanted by the customer. The entrepreneurial activities are not necessarily related to

The Innovation Process

a single person, the entrepreneur, but they can be part of a business process that is executed by a team of players. The same holds for the creative part of the innovation process.

The word 'idea' also needs some elucidation. It covers both technical and commercial ideas. The word idea should not be replaced by invention. Inventing is the process of transforming an idea or discovery into a technical application that can be protected by patents or by keeping the know-how confidential as proprietary technical information. Inventing and innovating are separate processes that may or may not overlap. An innovation process can start with an idea to use an invention in a new, specific way or in a new combination, or, alternatively, the innovative idea could come from the market place and initiate research work that leads to an invention.

The three stages

Company-based innovation is not a single stand-alone activity, but a business process to manage and optimise a complex, interacting range of multi-disciplinary activities, involving many parts of the company. Innovation needs careful management in order to be cost and time effective. Innovation is a supply chain process and the essence of effective innovation management is to understand that the path from idea to market goes through three distinct stages: generating and conceptualising ideas, developing and demonstrating the concepts, and investing to extract the value from the market place.

Stage 1   Idea generation and crystallisation
This stage is about generating ideas and developing them into a concept that can be communicated and discussed. The concept development phase includes identifying the link with the market and customer needs and the value to the stakeholders (company, customer, society). This is the creative phase, where ideas need to be nurtured, turned around, combined, taken apart and reassembled. Concepts will diverge first before they can converge into the preferred concept.

Stage 2   Development and demonstration
This stage develops the concept and assesses its feasibility in all aspects such as customer value, cost, technology, health and safety, ecology and economics, societal acceptability and demonstrating this with a working prototype. The more radical the new idea is, the more difficult this

stage tends to be. The development plans have to be adjusted regularly in view of new findings, budgets need adjustment and the outcome is uncertain till the last moment. Typically radical innovation has a low buy-in inside the company outside the innovation team, and few people in the rest of the company will see the value of the idea and many will argue that the money would be better spent on lower risk developments that will bring in money earlier and with better odds. Radical innovation projects do need a champion to guard the project through this high-uncertainty, high-risk stage. The purpose of this stage is to reduce the risks of the potential innovation to such an extent that the risk to invest money in the next phase becomes acceptable. This stage should finish with a working prototype and a sound conceptual business plan.

Stage 3   Investment and preparing for launch
This stage is the entrepreneurial stage, where major money may need to be committed. It involves creating the capabilities in the company that are needed to launch the product or service into the market. The capabilities include building up marketing and sales expertise, creating access to production and distribution facilities, and preparing a detailed business plan. This stage is similar to any project to bring a product or service to market, and can for the most part be managed with 'conventional' techniques. Only the interaction with the potential customers needs specific attention.

These three stages have universal validity and are independent of the business process or management system. They can be recognised throughout history, for instance in the pre-industrial era in the discovery of America by Columbus, the innovations by Edison in the industrial era and in Silicon Valley now.

There is some learning value in looking at the experiences of Columbus and Edison to identify the stages in their projects and to see how they managed the transitions, as done in Table 1.1.

Columbus is a good 'inventor' with a novel, creative idea. He struggles for many years to get a sponsor, but without a champion he cannot go to the next stage. His performance in stage 2 is not quite perfect, since he does not really check his theory and does not recognise his error. The entrepreneurial phase is the weakest part in this innovation; there is no business plan and Columbus is not an entrepreneur. Columbus is an explorer who performs best in the first part of the innovation process.

The Innovation Process

Table 1.1 Columbus, Edison and the three-stage model

	Columbus and the discovery of America	Edison and the innovation of electric illumination
Stage 1	**1476–1490** Columbus develops his own theory on the route to The Indies via the West, makes a proposal to King Ferdinand and finally gets it accepted.	**1878–1879** Edison starts experimenting with electric light bulbs and thinking about electricity distribution.
Stage 2	**1490–1492** Columbus sails West and finds The Indies at the expected position on the globe.	**1879–1880** Edison develops a light bulb with an acceptable life, a dynamo and a distribution system, and demonstrates the whole system at Menlo Park.
Stage 3	**1494–1500+** Columbus is appointed viceroy to start colonisation, but fails and is called back to Spain; the new colonies are ignored for a long time.	**1880–1882** Edison decides to finance the required investments himself and becomes an entrepreneur. Wall Street in New York is illuminated in 1882.

Edison is an outstanding inventor as well as innovator. He approaches a problem simultaneously from a technological perspective and with the application in the market in mind. He does not focus solely on the technical problems, but looks at the whole value chain. Edison is one of those remarkable figures who combines the skills of both the inventor and the entrepreneur and personally masters all the stages of innovation: the creative, the development and the commercialisation.

Columbus is the proverbial explorer with novel ideas, who likes to prove them, but has no interest in creating value out of them. He prefers to tackle the next idea. Edison is both inventor and entrepreneur and that combination is needed for successful innovation. It is the challenge of a good innovation management system to ensure that both the invention and the entrepreneurial aspects have their appropriate place and link up seamlessly.

These learning points show that the innovation process has some universal aspects that remain valid through time and they underpin the second law of innovation.

Law II The innovation process has three distinct stages: ideation, development and investment; the prime requirement for stage 1 is insight, for stage 2 a champion and for stage 3 an entrepreneur

The key requirements for stages 2 and 3 will not be surprising. The requirement for a champion in stage two, who pays the bill and offers 'protection' during this uncertain stage, is well known. The need for a risk taking, entrepreneurial approach in stage 3 is also obvious, but many people may have expected the word 'creativity' for stage 1. We have used the word insight rather than creativity. Of course creativity always is an important element in a novel idea, but it is the combination of creativity and analyses that transforms a 'novel idea' into the 'right idea'. Innovation needs insight; creativity by itself is not enough. The new idea needs to be grounded in deep understanding of its technical merits and the requirements of the market place. Creativity needs time and analyses to mature into insight.

Edison got to the right idea by moving from an electric light to electric illumination, based on his insightful understanding of the desired application in the marketplace. For Columbus, one could draw a similar observation: the idea to sail west would be the creative part, but the analyses that the western route was about equally long as the eastern one made it an insightful proposal. Unfortunately it is also true that when the analysis is wrong, the insight is wrong as well.

But the most important reason for preferring the word insight over creativity is that a truly grand idea always has an element of a vision; the intent and expectation that the idea is going to have an impact and make a difference. Columbus and Edison both had this vision. This difference between creativity and insight still holds today, as any innovation manager will know. Idea generation workshops with laymen only can generate many new and creative ideas, but these ideas have to be screened and nurtured by experts to create truly valuable ideas. In the end most insightful ideas come from professionals, although they may have been primed and stimulated by outsiders.

But besides the time-independent aspects of innovation, there are significant differences in the context in which innovation takes place through history. Columbus operated in the pre-industrial era and Edison in the industrial era. To arrive at an innovation model that fits today's business environment we have to understand the changes that have taken place over time.

THE PROCESS OF INNOVATION

It is tempting to think that if we learned from Edison and copied his approach we would have the start of a good innovation management system. But we live in a different era and the way innovation occurs has changed over time. Edison provides a perfect example of how innovation typically occurred in the industrial era, as Columbus does for the pre-industrial era. The world is now moving to the post-industrial era and the innovation process is changing with it. We have passed the point that optimisation of the classical innovation approach as used by Edison is the best option.

The classic innovation model

Today's innovation processes have their roots in the industrial revolution. Before the industrial era, innovation occurred mainly via trial and error – and serendipity must have played a major role. Of course there was good thinking and some experimentation, but the approach was not very systematic. The world and the worldview were, relatively speaking, static in nature and where there is little change there is little scope for innovation.

The industrial revolution brought the systematic application of science and technology to the innovation process. This created the classic approach to 'technology enabled innovation' based on scientific discoveries. Scientific discoveries can be converted into inventions in the technology domain and some of the inventions are developed further in the business domain and brought to market to become innovations that diffuse through society.

The result is a type of cascading overflow model of innovation, as illustrated in Figure 1.1. This approach has been extremely successful and has shaped modern industrial society. Many of the great innovations such as electricity, electronic communications and medicines have been developed in this way and started life in the laboratories for fundamental research.

Of course there are many and famous exceptions to this sequential model. The Wright brothers took to the air before science understood the principles of flying. But generally the innovation process was structured along the cascading, functional sequence with innovation starting in research. Many senior managers started their career in this mode of operation and still have the classic innovation process as the reference management model.

Insight in Innovation

discovery — science domain
invention — technology domain
innovation — business domain
diffusion — society domain

technology driven, sequential innovation

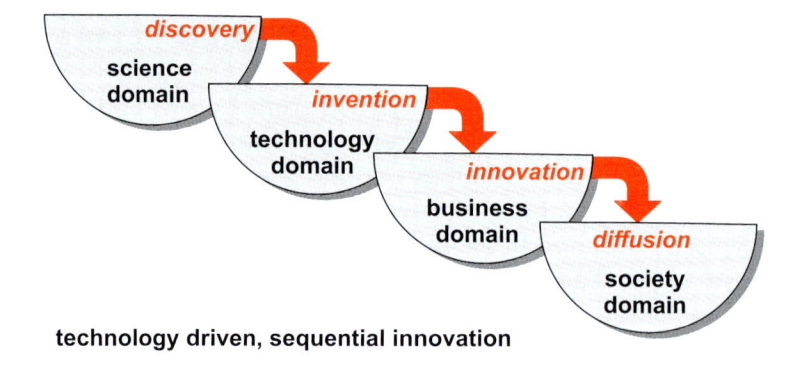

Figure 1.1   The classic, cascading model of innovation.

To arrive at a new model for the innovation process that fits the post-industrial era, we have to understand the changes that have occurred in the business environment and the way companies organise themselves. The main changes relate to the role of science and technology, the emergence of modern management techniques, the evolving market dynamics, the need for input from customers and other stakeholders, and the break-up of the innovation supply chain within large companies.

In the classic innovation model technology is the main driver for innovation. Technology is the source of novel ideas and new applications based on scientific discoveries that are 'pushed' into the market. The industrial innovation system was driven by science and technology, and this approach worked well as long as the new inventions tackled basic needs and supported economic growth in a low-income world. Innovation was by and large supply driven. Effectively society was waiting for all the new inventions and the fruits of science and technology were equated with progress.

In the rich part of the world this is no longer the case. The customer wants choice, but his choices are hard to predict at the start of the innovation process, and furthermore society wants to be aware of the risks and possible side effects. The supply-driven approach worked well in the early stages of the industrial era when the new products improved the basic needs of life and were almost automatically valued by the customers. Choice for the customer was less of an issue as the story of the black model-T Ford shows. A hundred years ago, a low cost car represented such a major improvement in mobility that for most people the fact that the car could only be delivered in black was

The Innovation Process

hardly an issue. But at today's level of wealth the customer wants a choice. Thus innovation that does not include the customer in the process is no longer fit for purpose; the risk is too high that the customer will not be willing to pay for it. The system has changed from being supply driven to demand driven, and from being steered by technological possibilities to being steered by customer needs. Whereas in the industrial era technology was the dominant agent for change and thus for innovation, in today's world there are two agents for change: technology and markets. In today's wealthy societies the markets are complex and sophisticated and have their own momentum independent of the technology used. Innovation is about creating links between technology and markets, and both sides have their own dynamics.

In the context of innovation, technology covers more than just science-based skills and applications, but also includes other business capabilities such as management, finance and marketing techniques. The most significant change in the post-industrial innovation process is the emergence of modern management techniques. Science and technology are no longer the dominant creators of wealth and added value in business; management techniques and business models have become significant value-adding mechanisms in their own right. In many mature industries technology is a commodity and not a differentiator or at least no longer the exclusive or dominant one. In industries such as aviation, shipping or refining, the business model is equally or more important for competitive advantage than the technology. This is not to say that science and technology are not important, but that the business model on how to extract value in the market place is the key to success. The 'how' you use it has become equally or more important than the 'what' you use.

A modern battleground for innovation provides a striking example. Distributed power is an emerging trend enabled by new technologies such as fuel cells, micro-turbines and Stirling engines [7]. Many companies anticipate this trend and expend a lot of effort in creating a competitive position. Still with the technology and the market needs emerging, most companies struggle to make money and some of the big players have left the arena before the competition really started. The core problem is the business model. The equipment by itself does not create enough value to enter the market; it has to be linked to a service. Success will depend on finding the right package that creates a compelling value proposition to the customer and a cost-effective way of delivering the package. The technology of distributed power is the enabling agent in the innovation, but the determining factor for success is the way it is used to add value to the customer and allow the company to extract value.

From a management perspective, a weakness of the traditional cascading innovation process is its sequential nature and associated lack of a feedback loop. Only after an innovation has entered the market and starts to diffuse can the effects on society at large become clear. When the innovation brings unexpected side effects, for instance on the environment, a long time will have passed before remedial action can be taken in this sequential process.

The time constant of the sequential cascading process is also too long for another reason. The development time is long compared to the dynamics in the market, and the long lead-time is particularly unattractive in mature markets with declining margins.

As a consequence the 'innovation supply chain' inside companies has broken up. Most companies have withdrawn from the science domain and leave fundamental research to universities and government sponsored research institutes. John Buckley [8] is quite succinct about this point and does not consider this withdrawal an option, but an absolute necessity. In his opinion company-based research is bad business. Research is the domain of academics; it is 'funded institutional curiosity'. Buckley's position may be somewhat extreme, but the fact is that most large companies have no or only a very limited effort in scientific research and can no longer depend on in-house science as a source for inventions and a starting point for innovation.

Having the whole innovation process in-house was more or less the standard in the industrial age. Research centres such as the Bell Laboratories produced both Nobel-prize winners as well as innovations that changed the face of the earth. Nowadays a company has to work with one or more external parties at one or more points in the innovation process.

But the break-up of the in-house innovation supply chain has gone further than abandoning fundamental research. Many companies have outsourced their development or manufacturing capacities and some innovative companies have hardly any in-house capabilities at all. This break-up of the in-house process has not occurred solely for economic reasons. Innovation has become so complex, competitive and multi-disciplinary that few companies can hold all the expertise in-house at world-class level. It is more efficient, and often the only option, to work with partners that already have the specific expertise and capabilities in one or more stages of the innovation process. Partners and strategic alliances are essential to reduce the risks as well as the 'time to market'.

The Innovation Process

Research is an important component in the innovation process and companies have changed the way they organise research. The time when research was a separate empire in the corporation, managed quite independently according to its own rules, has passed. In 'Third generation R&D' [9], one of the classic books on research management, the authors describe the three broad approaches to research management that companies have adopted.

> "Some firms leave resource allocation largely to R&D management. Many others frustrated by the difficulty of realizing payoffs from large R&D investments, have moved toward a more systematic second generation model: they subject R&D projects to the same kind of cost-benefit scrutiny applied to other business investments. In 3rd generation R&D, corporate, business, and R&D management must act as one to integrate corporate, business and R&D plans into a single action plan that optimally serves the near-, mid-, and long-term strategies of the company."

The message was that research needed to open up, abandon its insular position, and become integrated in the company strategy and innovation supply chain. In many companies innovation will have to go through a similar development and in the above paragraph the word R&D could more or less be replaced by innovation.

> "Some firms leave resource allocation for innovation largely to R&D management. Many others, frustrated by the difficulty of realising payoffs from large innovation investments, have moved toward a more systematic second generation model: they subject innovation projects to the same kind of cost-benefit scrutiny applied to other business investments. In 3rd generation innovation management, corporate, business, and R&D strategies must be integrated into an innovation plan that optimally serves the near-, mid-, and long-term strategies of the company."

The bridge-building innovation model

The contrast of the new mode of innovation with the classic cascading model is reflected in the bridge-building model. This model goes back to the roots of innovation described by Schumpeter [5] as finding new combinations between the technology domain and the markets. In this model the technology includes not only the science-based technologies, but all the capabilities that a company can muster, including management techniques and business models as well the partnerships and alliances that can be formed.

Inherent to this model is that innovation is less dependent on a single person, but is based on teamwork with a multi-disciplinary approach. Innovation does not necessarily start in the research laboratory or in the marketing department; it may begin in many places and be executed by diverse, multi-disciplinary innovation teams. This type of innovation is not supply driven, based on what the company likes to deliver, but starts with the needs in the market place and tries to make links with the strengths and capabilities of the company.

Law III Innovation is opportunity-driven; an opportunity is a value creating link between (potential) customer needs and (emerging) business and technological capabilities

As a result the supply chain for innovation is no longer the sequential series of steps from discovery to diffusion, or from science to society. Innovation starts as a scanning process at the interfaces between capabilities and markets to search for value-adding business opportunities and then to develop paths to realise them, see Figure 1.2. In this innovation model research is not at the beginning of the innovation chain, but at a later stage to develop the technologies that are

opportunity driven, linking innovation

*Technology: the total set of scientific and business techniques, processes and capabilities available to the company

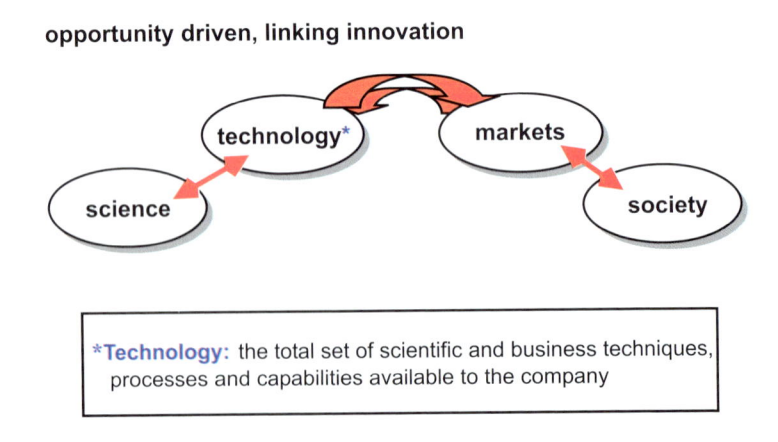

Figure 1.2 The bridge-building model of innovation.

The Innovation Process

required or, if need be, to underpin them with further scientific research. Innovation also includes the simultaneous development of the business model, creating new channels to the market or new links to society to gain acceptability.

At the same time the decision-making process in innovation is no longer made in a functional sequence. Many older innovation management models used a type of sequential filter process. First, ideas were tested for their technological feasibility. If this was demonstrated, the commercial feasibility was tested and finally the fit with company strategy was assessed before the new product was launched. Now these questions are all asked at the same time at each tollgate. The difference between the tollgates is not in the nature of the questions, but in the required degree of depth and sophistication of the answers.

Table 1.2 summarises the changes in the innovation environment from the industrial to the post-industrial era, and the associated changes inside companies and the innovation process.

Table 1.2 The two innovation models compared

	Classic cascading model	Bridge-building model
Innovation environment		
Cultural driver	Scientific curiosity	Business opportunity
Functional driver	Technology	Technology and market
Business driver	Supply	Demand
Company structure		
Organisation	Functional divisions	Business units
Innovation supply chain	Integrally in-house	Created with partners
Innovation process		
Development process	Sequential	Parallel
Assessment	Functional	Integral
Custodian	Research function	Business unit

The innovation processes through history

The shifts in the innovation processes throughout history can be compared at different levels. Not only the driving forces have

Table 1.3 The three innovation eras

	Pre-industrial	Industrial	Post-industrial
Business environment	Static markets	Emerging markets	Mature markets
Driving force	Philosophical concepts	Scientific curiosity	Business opportunity
Markets	New (meta-) physical worlds	New technological applications	New customer needs
Initiator	Explorer	Inventor	Innovator
Champion	King	Research director	CEO as entrepreneur
Development methodology	Trial and error	Systematic application of science and technology	Business process for creating new technology/market combinations
Management model	Royal edict and monopolies	Technology push and market pull	Innovation supply chain management

shifted over time, but the stakeholders, the nature of the markets and the key players have also changed. Table 1.3 summarises how the innovation system has changed over time at these various levels.

The table may give the impression that in each era an old approach to innovation is replaced by a new one. It is more appropriate to say that the old approach loses its dominance, but remains a valid option for a specific set of conditions. The classic cascading model fits best in an environment with emerging technologies and emerging markets, whereas the bridge-building model operates best in a mature environment. Successful innovation has cultural components, both societal and company based. The choice between the different approaches will vary with the maturity of the industry and the relative importance of research and technology in the innovation process. The trial-and-error method, pre-eminent in the pre-industrial era, still has its role to play and is used successfully by many a garage-entrepreneur. The supply-driven model can still be the preferred innovation mode in emerging economies or in periods when a new technology creates new markets, such as occurred in the early years of the Internet.

The Innovation Process

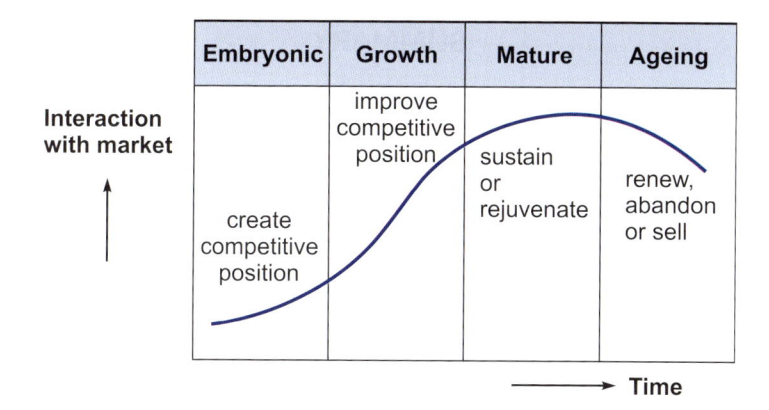

Figure 1.3 The role of innovation in the company lifecycle.

For a company or industry one could argue that the more mature it is in its lifecycle, the more the bridge-building model is applicable. Figure 1.3 gives the well-known S-curve for a company or industry with the four stages of development and the associated response to the market. If the markets are mature innovation starts in the market place and in undeveloped markets innovation can also start with new capabilities.

It goes without saying that the history of innovation is a bit more complex than a simple three period representation. More sophisticated books on the historic development of companies provide a richer backdrop than given here. Micklethwait and Wooldridge [10] identify seven stages in the history of business, from the merchants in the Middle East 3000 BC to the global multinationals of the last century. In line with this, one would expect that a seven-way historic division would be richer and more accurate. But the reduction to three stages makes it simpler to give an insight into the changes in the innovation process and the consequent changes in the management model.

Innovation is a much more complex process than it used to be. It is still about finding new combinations as Schumpeter defined it, but the new combinations operate in a more complex environment, it involves more stakeholders and requires input from external partners. Managing this complex system effectively, with the right balance of flexibility to give space for the creative idea and discipline to meet company objectives, is the key to successful innovation.

SUMMARY

1. Innovation is the business process for creating new and insightful ideas and bringing them successfully to market. The inspirational part of innovation is important, but the essence of effective innovation is managing the supply chain from idea to market efficiently.

2. The innovation business process has three fundamental stages.
   - idea generation and crystallisation
   - development and demonstration
   - investing and preparing for launch

   Edison was a great innovator who mastered the art of all three stages, but this combination is very rare and more typically each stage will need teams with different players.

3. The process for innovation has changed throughout history as a result of changes in the business environment. The main factors for change have been the changing role of science and technology, the emergence of modern management and business techniques, the evolving sophistication of the markets, and the changes in the innovation supply chain within large companies.

4. There are three main innovation process models developed over time in different historic eras, but these remain valid for specific conditions.
   - Pre-industrial era: The trial-and-error approach for the single inventor or entrepreneur and his 'dream'. The trial-and-error innovation model is a high-risk, unique opportunity approach that has its role in all times.
   - Industrial era: The classic cascading model driven by scientific curiosity with a sequential development process based on the systematic application of science and technology. The model works well in emerging markets.
   - Post-industrial era: The opportunity-driven bridge-building model with simultaneous development of technology and the business model. It is a systematically managed business process to optimise the use of a company's capabilities to achieve strategic objectives.

– 2 –
The Innovation Spectrum

CLASSES OF INNOVATION

Innovation covers a wide spectrum of business opportunities based on new technology/market combinations, ranging from minor improvements to an existing product with a time to market of a few months to a complete new business based on breakthrough technology taking more than a decade to develop. It is reasonable to expect that the innovation business processes may differ for these two extremes, and that factors such as the type of product, the company management style, the dynamics of the market, or the maturity of the industry can have an effect on the way the innovation process needs to be managed.

Many classifications of types of innovation have been proposed in the literature; Gaynor [11] gives a comprehensive classification and refers to it as the standard categories of innovation.

Table 2.1 The standard classification of innovations

	Service	Process	Product	Component	Material
Incremental	Modifications, refinements, enhancements, simplification			
Discontinuous	Obsoletes technologies, processes, and people			
Architectural	Changes core design concept to new architecture			
Systems	Dominated by societal and government regulations			
Radical	Develops into major new business or spawns an industry			
Disruptive	Brings the user a new value proposition			
Breakthrough	Moments in history that set the stage for the future			

For our purposes we need a classification that discriminates between the types of business process required for the class of innovation.

Two types of innovation

The classification as given in table 2.1 is with 35 categories very comprehensive, but not simple to use. The distinction between the categories is not always clear and Gaynor suggests reducing the categories to three only:

Incremental: Improvements to a current product* or class of products
New-to-the-Market: Novel replacements, including new to society products
Breakthrough: Changes the particular business or develops a new industry

This three-way classification is based on the type of product and the degree of novelty in the innovation and this translates into complexity and chances of success. This type of classification is quite common in the literature, and in particular the degree of novelty of the applied technology is often used; for instance, Tidd [12] uses the classification incremental, radical and transformation.

The same three-way classification is given by Arthur D. Little (ADL), but is based on consumer perception rather than technology [13]. According to ADL the typical customer responses for incremental, radical and breakthrough innovation are respectively "I know it's better", "I think it's better", and "Do I really need this". This sequence of responses may appear surprising. A natural sequence to the first two responses would be something like 'I hope it is better', indicating a further reduction of understanding by the customer, but suddenly the response expresses doubt about the necessity of the innovation. However, as we will see below, this difference in response between breakthrough innovation and the other types of innovation is real and significant.

---

*Throughout the book '**product**' refers to 'products, processes or services'.

The Innovation Spectrum

From the point of view of managing the innovation business process, classifications based on the type of product delivered or technology used are not easy to use, because they do not indicate whether the management process will have to be different. A better approach is not based on the technology or the product, but on the customer. A useful approach for classifying innovation based on customer response starts with the well-known Ansoff [14] matrix for classifying research. In this matrix as given in Figure 2.1, one axis gives the position of the target market place, the other the position of the technology in the research project. This well-known four-box matrix gives an indication of the degree of risk associated with a particular research project. The risk is based on the complexity of the innovation and the likelihood of customer acceptance.

The same matrix can be used for innovation projects, but for innovation management the classification reduces to only two possible types of innovation: **'inside-the-box'** and **'outside-the-box'**. Figure 2.2 illustrates this classification. An innovative idea can be outside-the-box because the customer is new to the company, or the product is completely new to the customer, or both, the so-called 'white space'. The consequence of these new-to-the-company aspects is that it will not be clear from the start of the project who will bring the innovation to the market or how it will be done. This knowledge is the essential difference between the two types of innovation, because the business processes required will differ depending on whether the route to the market is known a priori or not.

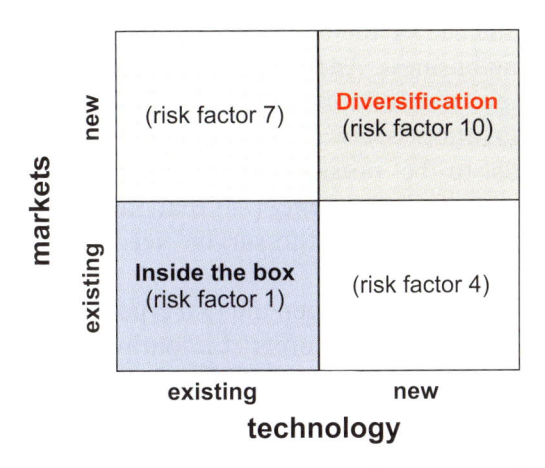

Figure 2.1 The Ansoff risk matrix for research projects.

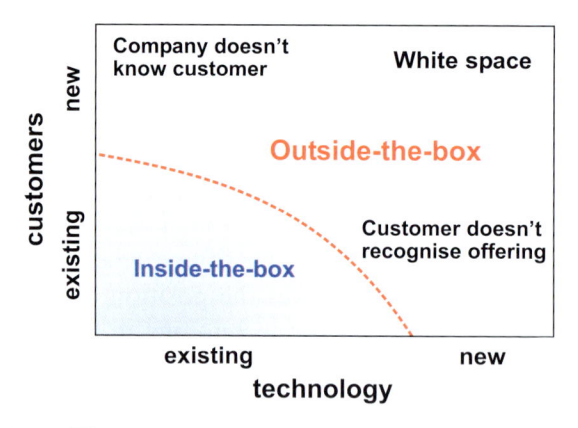

Figure 2.2 The management spectrum for innovation.

'Outside-the-box' innovation aims to change the rules of the game or to create a new paradigm, and is often based on a breakthrough technology or a new business concept. Outside-the-box innovation is also called revolutionary, radical, breakout or adaptive innovation.

'Inside-the-box' innovation, also called evolutionary, competitive or incremental innovation, aims to improve a product or refresh the competitive position of a business.

The various alternative names for the two types of innovation are not completely identical. Adaptive innovation is a term adopted from the evolution theory and refers to radical innovation in response to changes in the environment. It aims to improve the chance of survival of a company in a turbulent transition period in the business environment. It improves the resilience of the company by creating new businesses or new ways of doing business. Adaptive innovation is the opposite of competitive innovation, which improves the position of existing businesses. Competitive innovation is always inside-the-box innovation; adaptive innovation needs to be outside-the-box. Revolutionary innovation carries the message of Gary Hamel [15] that the age of revolution has arrived and companies that want to survive need to reinvent themselves continually.

The essential differences between the two types of innovation concern knowledge of the potential customer, the pathway to the market and the identity of the entrepreneur – the risk-taker who will bring the idea to market.

For inside-the-box innovations the customer is known and the way he may respond can be assessed, because the customer knows the

The Innovation Spectrum

existing product and only certain features of the product are changed. The marketing function plays the role of the entrepreneur and brings the idea to market via established methods and channels.

For outside-the-box innovation the customer or the route to the customer is not known, because the product cannot be demonstrated and can be described only in terms of desirable functionalities. As a consequence, the customer cannot really know whether he will like the innovation or not. Furthermore, the final customer can be totally different from the one the inventor had in mind originally. The risks associated with the innovation are thus high. A specific issue is that the usual channels to market can probably not be used. This means that the marketing department may not be interested in the product and an alternative entrepreneur will have to be found inside or outside the company.

Whether innovation is inside- or outside-the-box differs from company to company. What is a new way of doing business for one can be familiar ground to another, and an unknown customer for one company can be a preferred customer for another.

The distinction between inside- and outside-the-box innovation is fundamental, but this does not take away that within these categories there may be significant differences between companies in the execution of the management process.

Outside-the-box innovation

Inside-the-box innovation is incremental innovation based on existing or improved technology to support and develop the existing businesses. Incremental innovation has well defined customers and actors in the process. It is executed, managed and funded by a business unit and governed by unit business processes. Incremental innovation supports the product strategy of the business unit and is the responsibility of the business manager. This type of innovation is a 'must do' activity, required by any business and is managed according a business specific, internal business process. The business unit, marketing and research managers play this innovation game jointly. Inside-the-box innovation operates at product strategy level and is steered and funded by a business unit.

Outside-the-box innovation is game-changing innovation to create a step-change in the business for company strategy reasons or in response to radical changes in the business environment. This type of innovation stretches or goes beyond the existing business domains in the company.

Game-changing innovation is not so much a necessity, but one of the strategic options that a company can decide whether to use or not. This is not a game for the marketing or the research manager, but for the CEO. He needs to decide why, where and how much the company needs to change and why innovation is used as the preferred strategic tool to achieve the required changes. These changes can include the development of new business domains, new capabilities or new ways of doing business. Radical innovation is an alternative strategic option to mergers or acquisitions. By its nature, outside-the-box innovation is a corporate activity that is steered and funded centrally.

The intrinsic differences between the two types of innovation processes are the basis of the fourth law of innovation.

Law IV Innovation management distinguishes only two types of innovation: inside-the-box and outside-the-box, based on whether the pathway to the customer is known at the start

For a manager the essential difference between outside- and inside-the-box could also be summarised as 'mastering a new art' and 'meeting the agreed targets'. Both classes of innovations are investments for growth, but for the CEO outside-the-box innovation also has the element of an insurance premium to prepare the company for or protect against expected and unexpected changes in the business environment. Inside-the-box innovation is done to prosper, outside-the-box innovation aims at continuity and survival.

Typically, the risks associated with outside-the-box innovation are much higher than those with inside-the-box innovation; a factor of 5 to 10 is a normal range. But this characteristic, although helpful, is not the real discriminator with respect to the way an innovation project has to be managed. Also, the distinction as to whether the game-changing innovation represents something completely new or is only a substitution for an existing product can be important, but not for the management process. Furthermore the time horizon of outside-the-box innovation tends to be longer than for inside-the-box. Table 2.2 gives a number of characteristic differences between inside- and outside-the-box innovation.

None of these characteristics is a real discriminator, only the knowledge about the route to the customer is. However, in practice an innovation that strictly speaking is inside-the-box may be managed and funded as an outside-the-box innovation, for instance when the risks are high and the time horizon is very long.

The Innovation Spectrum

Table 2.2 Characteristics of the two types of innovation

Inside-the-box	Outside-the-box
• Incremental innovation	• Game-changing innovation
• Operates at product strategy level	• Operates at company strategy level
• Managed and funded by business unit	• Steered and funded corporately
• A 'must-do' activity	• Strategic option
• In-house development	• Developed with partners

The intrinsic differences between the two types of innovation affect all aspects of the innovation process, including the nature of the objectives, deliverables, management process or champion as indicated in Table 2.3.

Both inside- and outside-the-box innovation processes are staged processes, but the staging criteria are different. In general, strict staging

Table 2.3 Process features of the two types of innovation

	Outside-the-box	Inside-the-box
Deliverables	New business options	New customer value propositions
Positioning	Responding to a radical change in business environment or intending to create a paradigm change	Trying to create a competitive advantage or responding to competitive challenges
Company capabilities/ technologies applied	New	Existing
Sponsor/Champion	Corporate centre/CEO	Business unit/Product manager
Time horizon	2–10 years	1–5 years
Chance of success*	Low (<1:10)	Medium/high (>1:2)

*It is difficult and dangerous to give typical figures for the chances of success. The numbers very much depend on the type of industry and technology, and the way the innovation funnel is designed and managed. The figures given are not untypical for the ratio of approved ideas and market launches in mature industries.

Insight in Innovation

is less important for inside-the-box innovation. The reason for this is that the route from idea to market is well established and the players and their roles are well defined and described in the Quality Management (QM) manual for product development. The staging of the process is based on the deliverables required at the end of each stage and the management process can be designed with more stages or with less distinct tollgates to diffuse the staging.

The model developed by Cooper [16] provides a good example of a stage-gated business process for inside-the-box innovation. The model given in Figure 2.3 shows six stages that appear to be equal in weight. On closer examination, it appears that the development stage has been segmented into four sub-stages: detailed investigation, concept development, system development, and testing and validation. Sub-dividing the three basic innovation stages can be useful for better monitoring and control of progress, and usually the additional tollgates are positioned at the points when new monetary commitments have to be made.

It is good practice for the inside-the-box innovation process that it is carried out by integrated teams, including research and marketing staff, as over the years they will learn to understand each other's areas of expertise. For inside-the-box innovation the staging of the business process is less essential, and there is flexibility in the number of stages or the degree of separation. In fact Gaynor [11] finds the staging process restrictive for the innovation process and advocates de-staging it.

In contrast, for outside-the-box innovation the three stages are essential; each stage has to be managed in a specific style and requires different actors, and at each stage a different level of commitment is made. Game-changing innovation is a *'play with three acts'*.

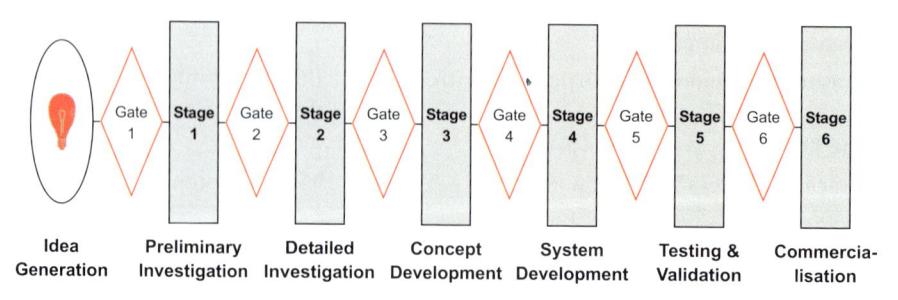

Figure 2.3 Stage-gated innovation business process.

The Innovation Spectrum

To appreciate the difference between the two business processes we will use an analogy as an illustration. Compare the innovation process with an engineering system for facilitating sailing upstream a fast flowing river. When the fall of the river increases a stronger motor may help, but at a certain drop the river needs to be staged and locks have to be installed to manage the current. At extreme height differences, even locks may not work and the flow will need to be controlled by dams. Cargo will have to change ship to travel further upstream or in innovation terms: the innovation may have to change company.

WAYS OF INNOVATION

From a management perspective the dominant distinction is the classification inside- or outside-the-box. But it is important to appreciate the difference in dynamics that can occur in the way the innovation is initiated and executed. Three basic modes are relevant.

- *Initiate innovation in the top-down or the bottom-up mode*
  In the bottom-up approach grassroots ideas are used to create an internal market for innovative ideas, whereas in the top-down approach management sets the direction.
- *Execute the innovation supply chain alone or with alliances*
  For inside-the-box innovation the all-alone option is often preferred, but for outside-the-box innovation there is usually no good alternative to doing it with partners
- *Be cautiously innovative in the comfort zone or seriously innovative in the impact zone*
  If innovation is not a core strategy, innovation in the comfort zone can be an attractive low-risk option as part of an acquisition strategy. If the company's innovation capability is a competitive advantage, innovation in the impact zone is the strategy of choice.

Top-down and bottom-up

Two approaches can be used to initiate innovation in a company: top-down or bottom-up. In the top-down approach company management defines the specific strategic objectives for the innovation effort, or identifies the problem areas that need remediation. In the bottom-up approach management only defines the domains of innovation, and it is left to the grassroots effort to develop attractive business opportunities for those business domains.

Both approaches have their pros and cons. The top-down approach gives more focus, simpler buy-in in the company, easier resource management, and can be more effective. However, the danger is that the screening process is less rigorous, because it is assumed that the ideas have support from the top.

Grassroots innovation has a better chance of generating out-of-the box ideas, but is always in danger of degenerating into a random process or hobby-ism. The bottom-up ideas can be screened more

GAMECHANGER – THE SHELL WAY TO INNOVATE

The Shell GameChanger story has been told already in various ways [18–20]. The GameChanger process was designed to stimulate outside-the-box, bottom-up innovation. A team of professionals was created to stimulate and coordinate the activities in the first stages of the innovation funnel. The aim was to harvest the creativity throughout the company by providing 'instant' support with a minimum of bureaucracy to any member of staff in the company with a good idea. Key features of the GameChanger process are:

- Easy and simple submission of ideas to a peer group, rather than to senior management, via the GameChanger website; the initial screening of ideas by 'horizontal filters' works out better and is more user-friendly.
- Response within days and, if an idea is accepted, a small budget is provided to develop the idea to a proposal with professional support if required.
- Frequent stage-gate meetings to support quick decisions and rapid progress.
- A special unit with access to venture capital was created to pick up ideas in the development stage and bring them to market.

Another approach has been adopted for top-down, outside-the-box innovation, the FRD (Fast Results Delivery) initiative. For big ideas, dedicated, global, multi-disciplinary teams are formed that have to deliver the final proposal to top management within one hundred days. The time process is highly structured and non-negotiable. The teams work in virtual mode most of the time and they can mobilise 'instant support' from experts and management throughout the Shell Group of Companies.

The Innovation Spectrum

effectively, but it is more difficult to maintain the momentum in the innovation funnel. Finding the right balance between top-down and bottom-up is the challenge for the innovation manager. Without steer from top management, the innovation effort can lose focus and relevance; without a sound grassroots effort, the creative power of the company may not be used optimally and the span of solutions may suffer.

The difference in mechanisms for initiating top-down and bottom-up innovation can be illustrated with the story of Shell's experience with the GameChanger initiative.

It is a special case if the CEO has the creative idea and puts him/herself in charge. This is an extreme case of top-down innovation with the typical problems and opportunities of top-down innovation magnified. If there is a single learning from the stories about the famous heroes of innovation it is that there is no standard recipe for success and each story about a CEO and his/her success has its special twist [17]. Each innovation follows its own tortuous development path and needs its own approach to success.

Alliances or all-alone

A significant difference between inside- and outside-the-box innovation is in the execution of the innovation process. Inside-the-box innovation tends to be in-house, whereas outside-the-box innovation needs external support. Very few companies have all the competencies and capabilities in-house to execute the whole supply chain for outside-the-box innovation. Thus external input is required in one or more areas, be it generating new ideas, creating access to technologies or prototyping the new product. Inside-the-box innovation typically operates as a series of incremental improvements over time and usually all the required competencies are developed in-house in the process or can be accessed quickly via well-established routes.

The main reason for executing incremental innovation in-house is confidentiality. The output from the inside-the-box innovation supply chain tends to be small changes to existing products that are in direct competition in the market with similar products. Copying of the innovation can be relatively simple and this creates a serious risk, particularly when being first in the market is important. Cooperation with third parties can be the source of leaks that are hard to control, but potential contamination is also an issue. Uncontrolled inflow of confidential information into one's own area of expertise is highly undesirable and can lead to expensive claims. There are also external reasons supporting in-house development. Finding a partner can be problematic, because most companies with the right expertise will be competitors or would-be competitors. However, outsourcing certain parts of the innovation chain under contract to specialist companies or consultants can be part of in-house development.

For outside-the-box innovation the situation is different, because potential partners with the complementary expertise will tend to be in different markets. They may have affinity with the potential market for the new idea, and as a consequence also share the interest to explore and develop the innovation. Although leakage of confidential information and contamination may still be areas of concern, they are less critical because the cooperation will include a controlled exchange of confidential information and the information from each side on its own cannot be used as such. The value is in combining the information and the gain in time to market and the leverage by the new competencies are key advantages. Leverage is more important than leakage.

Finding the right partners in innovation is not easy, because interests are conflicting and creating a win-win situation is not simple. The partner should bring complementary capabilities, but preferably also have little or no overlap in other areas of competence or business domains to avoid contamination issues or potential conflicts of interest. But the further removed the partner is in business terms, the more difficult cooperation may become in terms of communication. It can take a long time before the partners learn to understand each other's language; the same words can have different meaning in the two companies, development processes are not aligned and decisions are made at different points or against different criteria. It needs continuous vigilance to maintain a win-win situation with an attractive benefit/cost ratio for both parties. Both sides must have confidence in the opposite party and develop a mutual trust that both sides will stay in the game and play it fairly. This is a time-consuming process that needs to be based on a long-term advantage for both parties.

Openness to the world outside the own business is a characteristic feature of outside-the-box innovation. Innovation is carried out in an open environment, including networking with other centres of excellence, working jointly with strategic partners and in dialogue with society. In the pre-competitive stage of development the dialogue with society often takes the form of a request for governmental support. The granting of a subsidy can be seen as a positive judgement that the innovation has the potential to bring benefits to society at large.

The Innovation Spectrum

However, this open environment for innovation has its own set of characteristic risks.

In general, alliances in innovation are high-risk ventures that can go wrong for a wide range of reasons. The underlying causes are that the innovation process is volatile and the outcome uncertain. The parties in the alliance can differ in appreciation of the balance of risks and rewards at one or more points during the development, and any of these points can be the reason that the cooperation is discontinued.

However, at the end of the day, there is not much choice. In-house technical development of the innovation may not be possible due to lack of resources or capabilities or may be economically unattractive because of the high cost or the longer time to market. The need to create alliances is expressed by the fifth law of innovation.

Law V Outside-the-box innovation requires external partners with complementary capabilities to find and develop the best route to the customer

This law also points at a characteristic difference between the industrial and post-industrial innovation process. In the industrial era the preferred way to carry out innovation was in a protective environment. Research centres were located in the green, far away from the business to let the scientist think freely and undisturbed. It was considered important to first make the invention and to properly protect it with patents before the innovation process was started.

In the post-industrial era, researchers need and desire to work closely with the commercial people to explore the road to the customer jointly. Research and business centres are physically close or virtually connected and research is no longer an insular activity, but an integral part of the business strategy and development.

Innovation zones

In general the value of the innovation portfolio will not increase proportionally with the effort. At low levels the innovation projects have little in common and effectively they operate on a stand-alone basis with little interaction or sharing of capabilities. Thus each new project attracts relatively high cost for building up new competencies. Only when the total effort goes beyond a certain critical mass that can

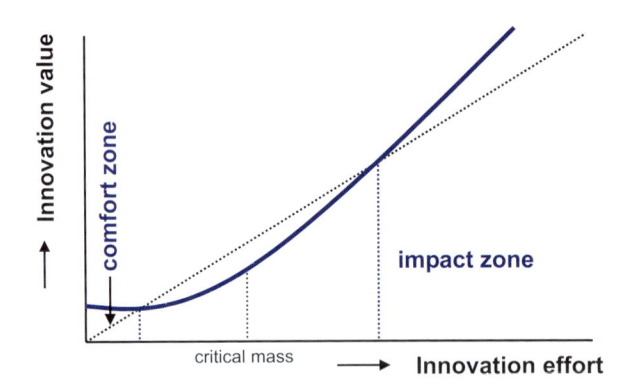

Figure 2.4 The two innovation zones.

support a balanced set of core competencies can value be added with relatively modest additional effort. Consequently there are two options for participating in radical innovation as illustrated in Figure 2.4.

The comfort zone
In this low-effort, low-risk approach the total innovation effort is below critical mass and the impact on the bottom-line of the company is small, even if the benefit/cost ratio of the effort is sound. The risks to the company are also limited. However, the overall value can be very attractive, because an important part of the value is intangible and derives from opening the windows to the outside world and the perception of being involved in innovation. The minimum requirement is that the innovation effort should be sufficiently high to keep the networking effective and the PR story credible.

This innovation strategy can be very successful in combination with a strategic acquisition effort. The innovation effort can lead to good opportunities for acquisition and provide expertise for smart buying.

The impact zone
In this high-risk, high-reward approach the innovation effort is beyond its critical mass and the results have a significant impact on the bottom line. Operating beyond the critical innovation mass means that there is strong interaction between the projects in the

The Innovation Spectrum

portfolio. Projects can have conflicting interests, but will also serve as platforms to generate new ideas and opportunities for innovation.

To operate in this zone a company needs a comprehensive range of capabilities and alliances that are tailored to meet the company innovation strategy. In the impact zone the innovation effort is so large that it affects the culture and the way of working in the company. The company will develop an innovation culture and company success depends on its success in innovation. If the innovation is well managed, such a company can combine high growth and profitability, as well as excellent resilience for continuity.

INNOVATION AND TRANSFORMATION

Usually discussions on innovation are focused on innovation for developing new products for external customers. However, in any company a lot of innovation is going on almost continuously to improve the way business is conducted. This can range from straightforward cost reduction measures to completely re-designing the business. This is innovation for internal customers. The customers can include company management, business unit management or other business units – for instance, improving the distribution system in support of the retail division. We will call this type of innovation for internal customers '**internal innovation**', in contrast to '**external innovation**' for external customers. This recognition of 'internal innovation' can be represented in a four-box matrix to represent the spectrum of innovation, with inside and outside innovation on one axis and internal and external customers on the other. Figure 2.5 represents the innovation spectrum with the GameChanger box indicating radical innovation for 'real' customers.

Internal innovation

Internal innovation processes are not usually described as such, but called, for instance, change process, reorganisation, re-engineering or redesign of the business process. Although the internal and external innovation processes have different names, similar laws and rules pertain, and only the customers and customer interests differ. External innovation tends to operate on 'adding value', internal innovation on 'reducing cost'.

Insight in Innovation

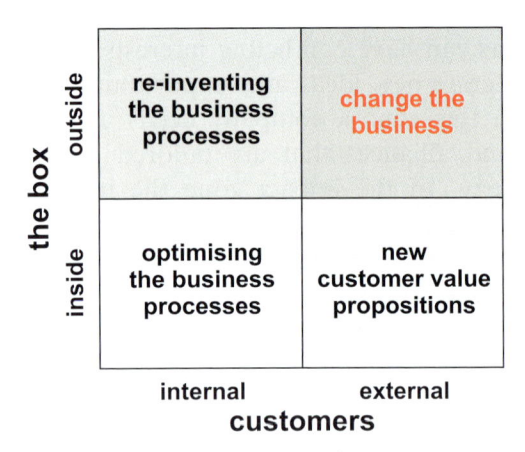

```
         outside  ┌─────────────┬─────────────┐
              │   │ re-inventing│ change the  │
              │   │ the business│ business    │
     the box │   │ processes   │             │
              │   ├─────────────┼─────────────┤
              │   │ optimising  │ new         │
              │   │ the business│ customer    │
         inside   │ processes   │ value       │
                  │             │ propositions│
                  └─────────────┴─────────────┘
                    internal      external
                         customers
```

Figure 2.5 The innovation spectrum.

Internal innovation is very much related to organisational change. Radical innovation and organisational transformation have much in common. Radical innovation always leads to change, and transformation of a business needs radical innovation. It is a fortuitous coincidence (or a logical consequence?) that the management processes for innovation and for change show clear parallels. Table 2.4 gives a few characteristics for internal innovation or organisational changes, both for gradual, inside-the-box and radical, outside-the-box changes.

Table 2.4 Characteristics of internal innovation (or organisational change processes)

	Outside-the-box	Inside-the-box
Deliverables	Creating new modes of operation for the business	Reducing the cost of doing business
Positioning	Responding to a change in business environment or intending to create a paradigm change	Responding to competitive challenges or trying to create a competitive advantage
Change	Transformation of the organisation and business processes	Improved business processes and operational efficiency
Process owner	CEO	Business unit manager

The Innovation Spectrum

Just like innovation, transformation of an organisation or company is also a process with three stages that have to be managed both as a whole, as well as with differentiation in each stage [21], just like outside-the-box innovation. This does not imply that both processes are the same, but that there are strong similarities between the dynamics of the two processes, and in the way they need to be managed.

Table 2.5 shows the stages and tollgates for game-changing innovation with the comparable phases of a radical transformation process expressed in the appropriate jargon.

The link between the radical innovation and radical transformation that is required to reinvent a business is shown pointedly in the story of the origin of Shell Global Solutions [22].

The changes at Shell have been recorded in business literature several times, most of the times as a transformation process [23,24]. However, there is value in portraying the formation of Shell Global Solutions as both a radical innovation and a transformation process executed in parallel. To illustrate this point, we will identify the three stages in the formation of Shell Global Solutions for both processes see Table 2.6 on page 40.

It is striking to see the similarity in dynamics of the two processes. They both follow their own internal logic, but are linked by the people involved and the underlying shared objectives. It is not easy to manage both these processes successfully and simultaneously. Both processes are complex and can change direction over time. There will be points in time when the direction as to where to go is not clear, but the leadership still has to lead and prevent the existing business

Table 2.5 Innovation and transformation are twin processes

	Outside-the-box innovation	Outside-the-box transformation
Stage 1	Creating a new idea	Letting go of the old concept
Transition	"*Find a Champion*"	"*Take time to mourn*"
Stage 2	Demonstrate the new idea	Adopt the new concept
Transition	"*Crossing the Valley of Death*"	"*Crossing the river*"
Stage 3	Prepare for launch	Launch new organisation

Insight in Innovation

THE ORIGIN OF SHELL GLOBAL SOLUTIONS

In 1996 the Shell Group of Companies went through a radical organisational change, restructuring itself from a country-based group of companies with 32 functions in the corporate centre into five global Group Businesses.

The downstream part of the oil activities became the Oil Products Business. The central research function was distributed across the five Businesses and in Oil Products (OP) it was combined with the Technical Service function to form OP-RTS (Oil Products-Research and Technical Services).

Life expectancy at birth was not high for OP-RTS, since it came with a heritage from a poor business era and a dissatisfied set of customers. Examples are:

- The Technical Service function had operated as a cost centre with an annual 'charge out' to its captive customers; the charge was based on the size of the customer rather than the service rendered and this system had become very unpopular.
- Certain parts of the OP business were not very profitable and were seen as opportunities for divestment. The need for R&D and services to support new investments or optimise existing facilities was considered to be well below the prevailing size of OP-RTS.
- Corporate centre had sponsored a significant amount of fundamental research and OP-RTS inherited a fair share of this research effort, but without a sponsor to pay for it.

This situation was not unusual at the time; many companies went through similar restructuring processes to reduce costs and improve effectiveness. The situation for OP-RTS was special in so far that the new challenge came completely unexpectedly to most staff. The Technical Service function had just completed a successful internal re-organisation and staff thought that no further change would be needed, because they were already lean and mean. It came as a shock, and almost as an unfair punishment, that 'suddenly' they had to face another round of staff reductions and find a way to survive in a business with customers that had a very negative opinion on the value of their services. It moved OP-RTS into a deep crisis, with depression, denial and chaos.

However, against all odds, within a few years OP-RTS turned itself around from a threatened cost centre to a very successful commercial knowledge company and became an example of 'survival of the fittest'. A few success factors are worth highlighting.

- A new business model was created that put the customer first, would deliver services only on request and against value rather than cost,

The Innovation Spectrum

THE ORIGIN OF SHELL GLOBAL SOLUTIONS (CONTINUED)
---

and with a flat organisation that made the basic units bottom-line responsible.
- Customers outside Shell would be as valuable as internal customers.
- The new organisation could offer a unique value proposition to its customers that none of its new competitors could match or imitate. It could offer integrated services along the whole value chain by staff that had actual hands-on experience in all aspects, ranging from research to refinery management.
- A new vision was developed for the role of technology and the value of know how and services.

The new business model was implemented in a fundamental and all-encompassing change process. Everybody was made responsible for the solution, and contributing to the solution was a necessary condition for survival. Proper time and effort was spent to make all staff aware of which phase in the change process that they were in, and advice and support were given on how to live through it emotionally.

Staff and leaders were top professionals who were committed to the business. There was no running away to safety or other escapist behaviour. Everybody shared in the losses as well as in the profits in a transparent and equitable way.

---

from falling. A strong leader with a clear vision is very important, but there is more needed than just this to be successful. The whole organisation has to move from one comfort zone to another by learning to understand and internalising the new vision and the new way of doing business.

In a way, Shell Global Solutions was in an advantageous position to carry out this process successfully. Less than five years earlier, the organisation had undergone a deep re-organisation to create a step change in effectiveness and professionalism. That process took several years and almost had failed when the direction of the change was lost in the complexity of the many initiatives. But in the end the change was very successful and the leadership had gained a lot of experience with major change processes, understood the dynamics and pitfalls, and — most importantly — was not afraid to go through it again. As a result, the second change process was managed almost flawlessly. Even the deep crisis in morale in

Table 2.6 The three stages in the formation of Shell Global Solutions

	Innovation	Transformation
Stage 1	**1996** The new business model for OP-RTS as a commercial consultancy developed	**1996** The old Shell (way of doing business) has gone; OP-RTS as is, will not survive in the new Shell. Significant downsizing and change on the cards
Transition	*Corporate support for new business model*	*Organic 'action learning' to change soft negative feelings of staff to hard business commitment*
Stage 2	**1997** Restructuring of OP-RTS from a functional into a commercial organisation, and preparation of business plan and technology strategy	**1997** Birth of a new, flat organisation based on semi-autonomous business groups (BGs); a BG has bottom-line responsibility for a specific product group
Transition	*Shell Global Solutions staff take entrepreneurial responsibility for their own destiny*	*'Learning by doing'. All staff involved in making the business plan for their BG and participating in public challenge sessions*
Stage 3	**1998** Endorsement of business plan and Technology Vision by corporate management and launch of Shell Global Solutions as a commercial company **2000** Celebration of successful start; Shell Global Solutions again is a great place to work	**1998** Launch of organisation with the new business model driven by the bottom line and customer satisfaction **2000** Celebration of successful start; Shell Global Solutions again is a great place to work

The Innovation Spectrum

1996 when the whole organisation had internalised the problem, but no solution was yet on the radar screen, was 'routinely' handled as part of the normal process.*

The process of redesigning the business was not recognised or managed as an innovation process. The process was seen as a transformation process with two major parts: the commercial part needed to create a new business model and the organisational part that had to create a new, commercial mindset in the staff. In retrospect, it was probably a good thing that the reinvention of OP-RTS was not managed as a game-changing innovation project. The business case for creating Shell Global Solutions as a commercial consultancy for the oil and gas industry was rather thin; effectively the new business was created as a result of the vision and entrepreneurship of one man and his team. The final success however came from the loyalty and professionalism of the staff. It is remarkable how few staff left voluntarily in 1996, when the ship was thought to be sinking. Empowerment of the staff proved to be more important in finding solutions than all the advice of leading change consultants.

The dynamics of innovation and transformation

Although the processes for radical innovation and transformation are similar, the dynamics of the two processes have their own momentum and logic, and at certain points will be out of phase. The emotional dynamics of the innovation process tend to be one step ahead of the transformation. For instance, when company management, as the innovation team, are on a high when the new business model takes shape on the drawing board, the rest of the organization can still be on a

---

*The 1992 reorganisation also created a grave danger. This re-engineering exercise was done for internal, functional reasons and had been very successful in improving morale and professionalism of staff. However, it simultaneously reduced the receptiveness of staff and management to external signals that the business was changing and different types of services were required. The organisation was focused on doing things right, rather than on doing the right thing. The 1992 exercise was 'competitive innovation' initiated internally; the 1996 exercise was 'adaptive innovation' enforced externally. Single-minded pursuit of the status of 'best in class' always carries the danger of missing critical changes in the environment, particularly when 'benchmarking' is used as the measuring tool.

Insight in Innovation

downward curve when staff realise that they have to let go of the old ways of doing business.

Figure 2.6 shows a stylistic example as given by Morrison on the variation of momentum during the innovation process, with two low points at the tollgates [25]. In general a successful project will show an overall increase in momentum with time with a few low points at various stages depending on progress. Typical low points are at the main hurdles, such as finding a champion and an entrepreneur. Figure 2.6 can be compared with the curve in Figure 2.7; this figure is taken from the Shell report on the creation of Shell Global Solutions and represents the judgement of the authors on the energy level in the organisation during the transition [22].

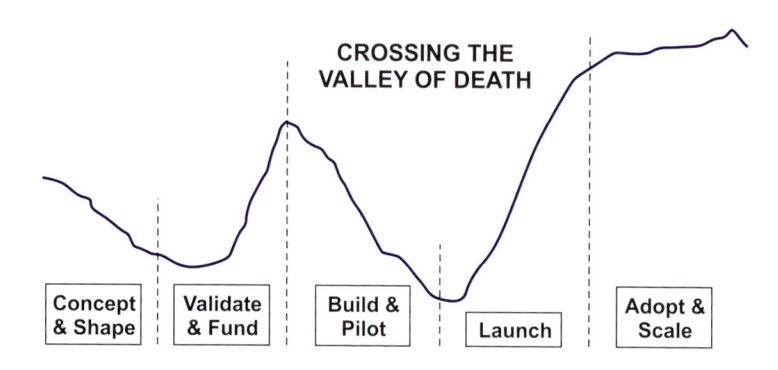

Figure 2.6 Dynamics in an innovation process.

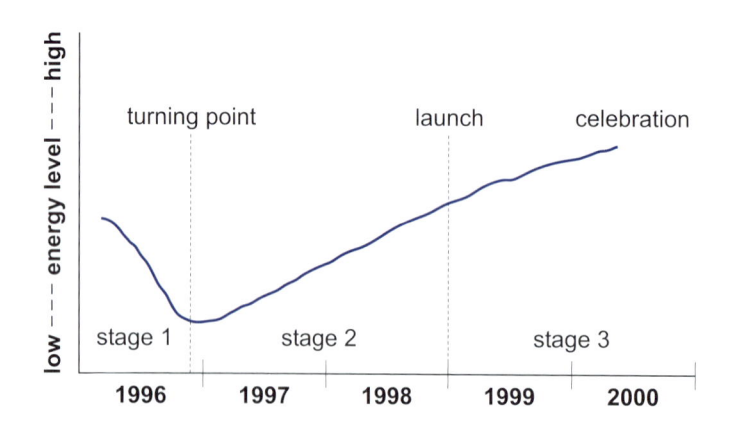

Figure 2.7 Dynamics in a transformation process.

The Innovation Spectrum

It is worth noting that the curve shows only one low point, but there could have been another one near the launch of the new entity when success was still very uncertain. It shows with how much confidence the new organisation launched itself into its new destiny.

Timing wise, the two processes, innovating the business and transforming the organisation, were very close in Shell Global Solutions, because the new business model was developed on the run during the change process. Thus the emotional levels of management and staff were not much out of phase, and both parties recognised each other's emotions. This was possibly one of the major success factors; it bonded the top-down and the bottom-up change processes at an emotional level. Most of the time the two processes will be further apart. Management will have analysed the business situation and designed the new model, before informing the organisation and starting the process of convincing staff and other stakeholders that a change is unavoidable. It is important that the organisation gets the time and space to get through its own dynamics, otherwise chances are that the internal conflicts will not be resolved [26] and business performance will suffer as a result.

SUMMARY

1. For innovation management only the distinction between inside- and outside-the-box innovation is relevant. Inside-the-box innovation is a 'must-do' activity and is managed and funded at business unit level and operates at the product strategy level. Outside-the-box innovation is a strategic tool that is managed and funded at corporate level.

2. Outside-the-box innovation needs external input. Although strategic alliances in innovation are risky and hard to manage, effectively there is no choice and the advantages outweigh the problems; leverage is more important than leakage.

3. Outside-the-box innovation needs to be both top-down as well as bottom-up to create momentum in the innovation process and make best use of the company's innovation capabilities.

4. Business innovation and organisational change processes are both staged processes and have similar dynamics; organisational transformation processes can be seen as outside-the-box innovation processes for internal customers.

5. Recognising and understanding the dynamics and the emotive aspects of transformation and innovation processes is a critical success factor.

– 3 –
Managing Innovation

INNOVATION IS A BUSINESS PROCESS

During the industrial era, the innovation processes in society and in a company were strikingly similar. The cascade from science to technology and from technology to business was repeated in the organisation of the company. Companies were organised functionally—with research, development, marketing and operational functions, and ideas cascaded through the organisation from research to the markets. In each function the innovation step was managed according to its internal, function-specific rules, and when the process step was completed the 'intermediate product' was handed over to the next function. This process worked well with science and technology as the 'invisible hand' to align the links in the supply chain, but at times the hand-over at the interfaces would lead to problems. Sometimes this happened simply due to poor communication, but more frequently it happened because the objectives of the functions were not aligned. Interface management was problematic in those organisations, because the functions were quite independent empires and alignment and conflict resolution could only occur at the top of the corporation.

In the post-industrial era the innovation process has become a quest to find and create new combinations between the technology and the business domains. Neither the search for new links nor the development or implementation can be done by separate units in sequential fashion; the innovation needs to be accomplished in a simultaneous, multi-functional approach. In the traditional innovation process the screening criteria for continuation came sequentially according to the progress through the functions: first scientific proof, then technical feasibility, followed by commercial attractiveness, strategic fit, operational acceptability, and so on. In the new model all the relevant questions are asked at each stage-gate and the analyses and conclusions from the previous stage are challenged in the next stage to test that the results are robust at all levels and from all perspectives.

The innovation supply chain

From a management perspective, it is useful to see the innovation process as a supply chain process with three main stages. Each stage needs to be operated efficiently and optimised in its own right, and at the same time the process has to be steered and optimised as a whole. This overall management is the responsibility of the innovation manager. He does not manage the individual stages, but ensures that each stage delivers as agreed, that the tollgates between the stages operate properly, and that the total innovation output can meet the strategic objectives. Understanding the nature and the purpose of the stages is important for managing the process.

Stage 1   Idea generation and crystallisation
In this stage ideas are generated, captured and nurtured to a stage where the potential business opportunity can be communicated and assessed. The idea is developed to a conceptual level and quantification of cost and benefits is rough and patchy. The key objective is that both the potential value to the customer can be identified as well as a conceptual business plan on how this value can be extracted outlined.

Stage 2   Development and demonstrating
In this phase the missing technologies and capabilities need to be developed and the feasibility of all aspects of the innovation—technical, environmental, production, logistics, customer response—evaluated and demonstrated. This phase should end with a working prototype and a quantified business plan. The key objective of this stage is to reduce the technical and commercial risks associated with the project to a level that investments become acceptable.

Stage 3   Building capability and preparing for market launch
In this phase investments to bring the innovation to market have to be made. These include not only financial investments in facilities, but also investments in customer understanding, recruitment and training. Mostly this stage can be managed as a normal product development or an investment project, with a number of special requirements to allow for the novelty aspects. A key deliverable is a plan for a successful

Managing Innovation

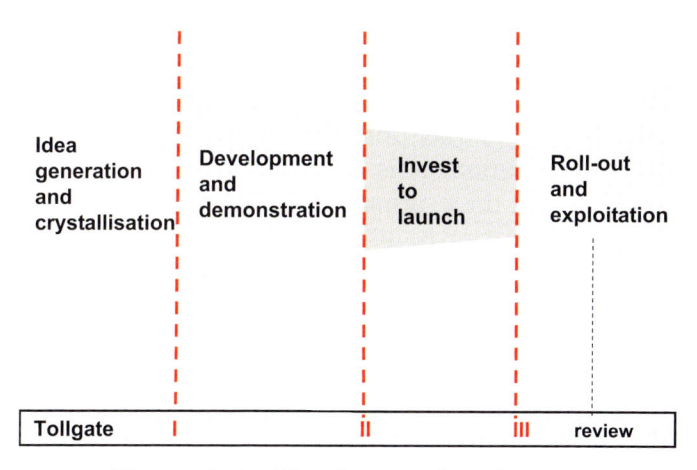

Figure 3.1 The innovation funnel.

launch of the product without exposing the reputation of the company at large.

The innovation process has often been described as an innovation funnel and a simple example is given in Figure 3.1.

The funnel concept conveys that during the process the number of innovation projects reduces as the poor ones are weeded out and the best ones get priority treatment. Managing the funnel has two key aspects:

- The funnel needs to be loaded at all stages
- Projects have to move forward through the funnel; backflow or stagnation means inefficiency

The innovation funnel has to be managed as a supply chain process for fairly basic reasons:

- The stages are sequential and interdependent
- Each stage has to be managed separately
- The overall process can be optimised with respect to use of resources and alignment of objectives.

The need for staging

An idea has to go through a transformation process to become an innovation, a process similar to the chrysalis of a cocoon to a butterfly.

Each stage in the transformation process has to be completed properly before the next stage can be entered.

Cooper [16] is considered the inventor of the stage-gated process for innovation. He developed a seven-stage model and applied it to accelerate the time-to-market for new, inside-the-box products. For outside-the-box innovation, it is important to define the innovation process first with the three primary stages, and then identify where further sub-structuring can be useful. The tollgates between the sub-stages are of sub-ordinate importance requiring directional rather than stop-go decisions.

Staging of the innovation process is the natural thing to do and many proposals have been made, for instance Roberts and Frohman [27], Quinn [28], Van de Ven [29], Berkhout [30], Cooper [16] and Gaynor [11]. The essential difference between the three-stage model and the other models is that in this model the staging is not based on specific objectives that have to be delivered at the end of each stage, but primarily on functional and managerial requirements derived from the intrinsic characteristics of each stage. Of course, the three-stage model also has a set of specific deliverables for each stage, but this is not the prime parameter.

The tollgates between the stages control the flow of ideas and projects through the funnel. At each tollgate a considered 'stop-go' decision has to be made, not only on the basis of the merits of the project as such, but also with respect to the whole portfolio of innovation projects.

The next step in the staging process for innovation is to place the right type of custodians at the gates. For out-of-the box innovation, the custodians of the tollgates need to come from company management, because of the high risk and the strategic implications. The task of the custodians is to guard the passage of a project from one stage to the next.

At each tollgate the members of the project may need to be changed to reflect the change in objectives and activities, including the shift from creative chaos to control. A few key players, for instance the original creator of the idea, may have to discontinue their involvement, because the next stage needs a different set of competencies. If this change of staff is not done properly, the innovation may not be steered by business objectives, but may drift into 'hobby-land' or be continued even when its value is fading. On the other hand, the key to successful innovation is often personal drive and commitment of a single person or a dedicated team to continue against the odds.

Managing Innovation

It is important that at each stage there is drive, dedication and commitment.

Not all authors are convinced about the usefulness of gates. Gaynor [11] argues:

> *"the stage-gate approach may present a tidy way for making decisions, but that's not the way innovation occurs. The process is not linear, but during development may hit unexpected hurdles and have to go back to an earlier stage."*

Gaynor's challenge is important, because it requires us to define the managerial need for the gates carefully. The gates are needed to separate the stages and to prevent backflow of ideas. Ideas may change direction within a stage of the innovation process as new information comes in, but they should not backflow. This would confuse the whole decision-making process, as the criteria for assessment are different for each stage. If an idea appears to be infeasible after exploring all options, it should be abandoned. If there are valuable elements that could be used, these should be included in a new idea, which should journey through the innovation funnel again starting from the beginning.

STAGES AND TOLLGATES

Each stage of the innovation process is home to a number of sub-processes, each with its own internal dynamics and purposes. The collective purpose of these sub-processes is to develop the idea, often in parallel activities, to the point that it meets the requirements for passing through the next tollgate. Each of the sub-stages has its own granularity with specific features.

The stages

Stage 1
The idea stage has three distinct parts: generating, capturing and nurturing, and for each part management has to decide how to structure and play the game. Since creativity has to blossom in stage 1, the structure should be open and flexible. It should allow for some chaos

Insight in Innovation

and empower people by giving them a bit of freedom, time and money to think and explore. The word chaos might be misleading; in this context the word 'uncontrolled' may be better.

The generation step should include a diversity of creative processes that are appropriate for the variety of the required opportunities and accommodate the different parties that can give an input: parties within the company, external stakeholders, partners and customers. Variety of actors and idea generation mechanisms is important because creative problem solving needs diversity. Table 3.1 gives a range of possible actors and the type of input they can give into the ideation process.

Table 3.1 Ideation actors and mechanisms

Actors	Purpose
Grassroots	Capture ideas from the work floor throughout company
Experts	Scan and analyse trends in business and technology for discontinuities and opportunities
Management	Define domains and purpose of innovation, and identify focus areas and targets
Customers	Extract new needs from deep understanding of (emerging) customer values and translate these insights into business opportunities
External stakeholders	Use input from external parties as a check on relevance and value, and as nuclei for change and new opportunities
Partners and alliances	Extract new opportunities from the know-how and expertise brought in by the partners in innovation

There are numerous books and guidelines on how to structure or stimulate the creative processes in a company. Well-known examples are the Seven Sources for Innovative Opportunity given by Drucker,* or the 40 Triz Keys to Technical Innovation [31]. An extensive discussion on this topic does not fit in this book, but idea generation is an

---

*The Seven Sources are: The Unexpected, Incongruities, Process Need, Industry and Market Structures, Demographics, Changes in Perception and New Knowledge.

Managing Innovation

important feature that needs careful design. It reflects the company's innovation culture, it is the starting point of the company's innovation effort and if this part is not successful the whole innovation chain will suffer.

But creativity by itself is not enough. An important understanding is: 'creativity needs time and effort to become insight'. This statement sounds counter intuitive and in conflict with our image of the sudden flash of deep insight by the genius. The sudden flash phenomenon is important, but most of the times the flash of insight comes at the end of a long thinking process. That's why most good ideas still come from experts, although the experts may need to be triggered by outsiders.

Creativity workshops have their role to play in the generation stage, but creativity on the spot is rare. Most participants arrive at such a workshop with their creative idea firmly fixed in their head, and the workshop turns out to be a forum for promoting existing ideas. Exchanging the ideas is useful, but the more important part is the follow up to start working with the ideas in a structured and focused way.

The key roles of diversity and insight are expressed in the sixth law of innovation.

Law VI   Diversity is essential to stimulate creativity, creativity needs time and analyses to mature into insight, innovative ideas are based on insights

In Shell an almost obvious input into the idea generation process comes from the long-term business scenarios. They sketch possible future business environments and identify emerging trends in global issues, technology, and societal values and new ways of doing business. Scenarios transform facts and data on the future into insights about the future and that's why they are good input into the idea generation step.

An important activity in the idea generation stage is to ensure that adequate attention is paid to potentially disruptive events, such as technologies, business models or customer values. An effort should be made to ensure that, as a minimum, these potential disruptions are on the innovation radar screen and assessed regularly for threats and opportunities.

The idea capturing step has received much less attention in the literature than the idea generation stage, but in a way this step is more difficult. Generating ideas is 'easy', but recognising a good one is hard. In the idea generation step it is rare that an idea is formulated clearly and focused. It is more usual that the idea is quite foggy in terms of customer value and business opportunity, and this makes discriminating between good and bad ideas tricky. It can be hard to judge if the idea is intrinsically poor or just poorly described. In general, it is good not to be too tough in screening at the initial stages of the process, but to give ideas a chance to grow as long as they fit in the innovation domain, seem to have value, and do not violate the basic laws of nature, human rights or company business principles.

It is good practice to clearly separate the generation from the capturing stage, both in time and people. Screening of grassroots ideas should be done by staff close to, but different from, the idea generators; it should be a peer group rather than senior management. Senior management may crack down too early on ideas that are not bad, but just unripe and not yet ready to be assessed.

In the nurturing phase the idea should be developed to a stage where it can be assessed for its relevance to the business and whether it is ready and worthwhile entering stage 2 of the innovation funnel. There are two important processes in this phase: clustering and underpinning. Clustering is the process to enrich an idea by combining it with other ideas or parts of other ideas. This activity is likely to have started already in the previous phase, but should be completed until the idea has the richest possible content.

The underpinning phase should provide the best possible answers to the tollgate questions in order to pass to the next stage. The key requirements are: fit with the company strategy, value for the customer and a plan on how to execute the next stage. In many cases this part will take the form of a scouting study. Provided that the confidentiality of the idea can be safeguarded, it can be beneficial to outsource the customer-related part of the study; it provides objectivity and credibility to the outcome.

Stage 2
The second stage of the funnel is the development and demonstration phase, which can take many different formats and may last from a few weeks to many years. During the tollgate I assessment the direction and objectives for the project have been set and agreed and in stage 2 the task is to develop the project to the entrepreneurial stage. This

development has to be done simultaneously on two parallel tracks: business and technology. Stage 2 has several characteristic zones that are phased, but overlap to a large extent.

- Development and testing of the missing technologies and capabilities
- Assessing the feasibility of the idea in all it aspects: commerciality, production, logistics, channels, safety, ecology, societal acceptability, etc.
- Developing a sound and quantified business plan
- Demonstrate the idea with a working prototype.

Typically stage 2 developments will need the involvement of external partners, because one or more of the required capabilities will not be available in-house. Whereas it is rare for an innovation proposal to be pursued if none of the required competencies are available in-house, it is also unlikely for a radical innovation that all competencies are in-house. Thus creating cooperative arrangements with external parties either on the commercial or on the technology side is almost standard for radical innovation.

The main advantage of cooperation is the reduction in development risk and time. Developing the required capabilities in-house will typically take longer and may not be successful. The main disadvantage of cooperation is the time and effort it takes to create a win-win platform and maintaining that platform till the task has been completed. It can take a long time before the parties understand each other's language, have built sufficient mutual trust, and established procedures to safeguard intellectual property and avoid contamination.

It is not unusual for the innovative concept to change direction a few times in stage 2 for a variety of reasons. For instance, the technology turns out to have different features than expected, the strategic partner shifts position or the anticipated markets fade and new ones have to be sought. The development process in the second stage is an iterative process via which various paths to the market have to be explored and the best ones developed further. This volatility in the project makes it likely that a few sub-stages are needed to assess the continued viability of the idea. Typical staging points are a significant change in the anticipated customer base or the construction of a demonstration unit.

The volatility and uncertainty that envelops a project and the many hurdles that have to be taken can makes stage 2 a treacherous domain. Typically the project will need additional funding a number of times as a result of setbacks and changes in direction. On top of that, the buy-in is still limited in the company, the value is uncertain and competition for funds is tough. For that reason it is important for the project to have a champion at senior levels in stage 2, who can prevent a crisis from turning into a discontinuity. Even the best ideas go through dips and disappointments in this stage.

Stage 3
The third stage has several distinct steps: developing the final detailed business plan, actively managing the risks that cannot be eliminated, making the necessary investments in facilities and resources and preparing to launch. These activities should be managed by an entrepreneur who is willing to take the risk to bring the innovation to market. We will discuss the role of the entrepreneur more extensively in the next chapter. The entrepreneur can come from inside the company (marketing or business unit manager) or from outside (venture capitalist) or it can be a hybrid form (joint venture, strategic alliance).

Stage 3 may include major investments for manufacture and distribution and for preparing marketing and sales staff for the new product. Finding the risk capital for the investments is never easy. Innovation projects always carry a higher risk than repeat projects and this higher risk is not easily compensated by potentially higher rewards. The option of 'not doing' is always an attractive alternative.

The deliverable of this stage is a proposal to launch the new product for approval by the CEO and his management team as the guardians of the third and final tollgate. The proposal should include the launch plan, subsequent roll-out options and a post-implementation review.

The tollgates

There can be good reasons to create sub-stages in the business process for innovation, for instance to facilitate progress monitoring. In simple innovation processes it can be adequate and more efficient to diffuse the staging. But it is important to always recognise the three fundamental stages in innovation, because the transfer points or tollgates between the stages have an essential managerial functionality. The tollgates between the stages are there for three management reasons:

- To separate the functionalities in the innovation process: chaos from control, exploratory research from entrepreneurship, etc.

Managing Innovation

- To change or adjust the team and/or team leader
- To make the high level 'stop-go' decisions.

The tollgates are the points in the process for making key decisions on the way forward, with a choice between three basic options:

Continue to the next stage

It is good practice to make this decision only when all the objectives are fully and positively met. It is a bad but not uncommon practice to give a conditional green light pending the resolution of a few outstanding or unsatisfactory resolved issues.

Move the project from 'outside-the-box' to 'inside-the-box'

This is an important assessment. Often in the development phase innovation projects move closer to existing markets in order to improve their potential profitability, and with the concurrent reduced risk in development cost and time to market, the project may again fit in the innovation portfolio of one of the business units. If possible, and given that there is commitment in the business unit, this is usually the preferred option.

Stop the project

There are two good reasons to stop. The first one is that the idea is bad and it will never make money, and the second one is that the idea as such is good, but at the end of the day the company is not interested in playing that game.

For managing the tollgates it is important to have clear guidelines in place and that all actors understand and appreciate the criteria for passing the tollgate. Effective tollgate management throughout the innovation funnel is based on the use of the same criteria at each tollgate, but with an increasing request for detail and quantification. At the first tollgate the ideas will still be rather conceptual and the quantification patchy, but at the second tollgate all questions must be answered and quantification must be sufficiently detailed to allow risks to be assessed, preliminary business plans made and investment budgets prepared. Tollgate III requires a comprehensive and detailed business and launch plan.

The structure of a tollgate selection system is given in outline in Table 3.2. At each tollgate the same issue must be addressed in different wording to indicate the level of detail and evidence required. Although each company has to tailor-make its own rules in line with

Table 3.2 Generic tollgate criteria

Criterion	Tollgate I	Tollgate II	Tollgate III
Strategic criteria – fit with innovation strategy – value for customer			
Technological criteria – fit for purpose – IP position	Each criterion needs to be defined specifically for each gate in line with the required level of detail and quantification		
Economic criteria – value to company			
Portfolio criteria – risk profile			
Other criteria –			

its specific innovation strategy and management process, most criteria will be rather generic. A few quite typical criteria are given in Table 3.3.

Managing the tollgates is essential to keep each project healthy and contain the risk in the total innovation effort. Game-changing innovation is a risky business and, as a rule of thumb, less than one in four innovative ideas will successfully pass through the next gate. This ratio of 1:4 can be even worse depending on the type of industry and the way the innovation funnel is managed. Certain companies try to cast the net as wide as possible and in order to contain the cost, apply steep reductions in the beginning of the funnel. Pharmaceutical companies may operate at a ratio of well over 1:1,000 in the initial stage. Other companies like to manage a pipeline rather than a funnel by only adopting sure winners in the idea stage. The difference is determined by whether the main objective of innovation management is improving the odds or achieving a target number of new products. In the early stages the costs per project are relatively low and having more trials in those stages can be a good investment for increasing the flow through the innovation funnel.

It is difficult to judge the merits of an innovation project at the early tollgates. The strategic fit is still uncertain and the quantitative data have limited value. It is important not to rely solely on gut feel,

Managing Innovation

Table 3.3 Basic tollgate criteria

Tollgate I	Tollgate II	Tollgate III
• Benefit for a potential customer identified	• Customer value proposition and target customers defined	• Comprehensive and detailed business plan accepted
• Fit with company strategy	• Business plan developed and adopted by entrepreneur (inside or outside the company)	• Pre-launch targets achieved
• Intellectual property position evaluated	• Feasibility demonstrated with working prototype	• Action standards for launch and measures for roll-out defined
• Preliminary commercial and technological assessment	• Positive customer response	• All remaining risks actively managed
• Fits portfolio in terms of risk, timing and resource requirements	• Safety and ecology assessment	
	• Risks reduced to acceptable level that allows investments to be made	

but to make the decisions as rational as possible. It is useful to be aware of two common errors of judgement at the tollgates.

The first one is impatience when the 'big winner' does not appear soon enough and as a consequence the innovation effort is reduced prematurely. It is true that innovation needs the occasional star to make the whole innovation effort worthwhile, because most of the ideas will fail and most of the successful innovations will not do much better than breakeven for a long time. However, innovation needs continuity, not for each project as such, but for the whole portfolio in order to maintain the competences and improve the chances for success. It should be emphasised that the need for continuity refers only to the total innovation effort, for an individual project the willingness to stop is a key requirement for effective innovation management.

Another common error of judgement is the management conviction that '*I believe that I will recognise a big one, when I see one*'. This is a very dangerous assumption in outside-the-box innovation, particularly in the early stages of the funnel. Sometimes this is true, but most big innovations were recognised only as big, when they had become big. The basic assumption in game-changing innovation management can only be that '**the winner is unpredictable**'.

The tollgate criteria are important for keeping the assessments objective and consistent. However, there is a success factor that has a bias depending on the company culture. Two different philosophies can be adopted in managing the innovation funnel: the champion approach or the coaching approach.

In the champion approach the critical consideration is that no innovative idea will be successful without a champion and even the best ideas will die without a committed and energetic person to drive the idea and push it across the hurdles. In this philosophy the idea gets through the tollgates on the strength of the 'driver'. This approach makes sense in those cases when the driver has the capabilities for handling the idea from start to finish. The driver needs to have 'Edison-qualities' with the competencies of both the inventor and the entrepreneur. Although rare, this combination of capabilities is very valuable as it enhances the chances of success considerably.

In the coaching approach the idea generators and project managers get professional support to meet the specific requirements of out-of-the-box innovation. In this philosophy the idea gets through the funnel and passes the tollgate on its intrinsic merits and on the assumption that all the stages of the funnel are managed and executed professionally.

THE ROLES OF THE INNOVATION MANAGER

The management hierarchy for game-changing innovation includes three levels.

1. Company management sets the objectives and the budget for radical innovation. The CEO can have a specific role here, if the challenge is great and the objectives radical.
2. The innovation manager is the custodian of the innovation business process and responsible for achieving targets and objectives.

Managing Innovation

3. At the next level, a project manager picks-up the proposal at one tollgate and delivers it to the next stage with involvement of all relevant stakeholders in the development process.

Not all innovative companies have the position of innovation manager in their organisation. Of course there are many ways to manage and steer the whole innovation effort, but if one considers it as a supply chain process, the need for an innovation manager is evident. Each stage of the innovation supply chain has to be managed differently and in line with its specific characteristics, but the innovation manager has to maintain the integrity and effectiveness of the whole process. The logical position of the innovation manager is in the strategy department, where he/she will be responsible for achieving one of the strategic objectives of the company. This position is complementary to the manager of mergers and acquisitions or the manager of the corporate venture unit. The roles of the innovation manager can be described under three headings: manage the innovation supply chain, the links with company strategy and the innovation portfolio.

Managing the innovation supply chain

Keep the momentum
Maintaining the cohesion and momentum in the whole innovation supply chain is the first and prime challenge of the innovation manager. The momentum should stem from the drive of the innovators and the value of the projects, but part of this momentum must also be based on support from the top. Even the best ideas go through low points and the better the idea the higher the internal hurdles will be, because very good ideas will threaten existing businesses. Finding a champion who provides an umbrella for an idea through the critical development phases is part of the job of the innovation manager.

Keeping the momentum high is important. Ideas increase in value only when they move forward through the funnel, and innovation management is about keeping the projects moving through the funnel. If an idea does not move forward through the innovation funnel, it only creates cost not value.

It is important to pursue the objectives aggressively and ensure that the targets are delivered as agreed. Of course predicting the time and effort required to find a solution is hard and many developers are reluctant to commit to a time-line for the execution of the project. It should be mutually appreciated that most of the time the estimate will be wrong, but the effort it takes can be structured and planned, and plans can be adjusted and agreed. Innovation is not a black box process. Serendipity is a valuable element in research and innovation, but as Edison said: innovation is 99% perspiration.

Innovation needs both creativity and investments. Creativity needs openness, freshness and a willingness to say yes; it thrives on a bit of chaos. Investments need discipline, sound analyses and a willingness to say no. The chaos bit gives innovation a bad name among many CEOs who equate chaos to the absence of control and good management. They may still allow innovation in the company, but prefer to stay away from it and delegate it to the research or technology manager in the management team. Keeping the CEO involved in game-changing innovation is part of the job of the innovation manager.

Manage the tollgates
Another task of the innovation manager is to ensure that the appropriate decision makers properly control the tollgates between the stages. At each stage the innovative idea has to meet certain specific criteria and these have to be assessed by the custodians of the tollgate. For game-changing innovation this is always senior corporate management, possibly in varying combinations for different gates. At gate I the strategic fit with company objectives is the main criterion; no matter how valuable an idea might be it is no use pursuing it, if, in the end, the company is not willing to play the game in the market place. At gate II the main criterion is the willingness to invest and find the entrepreneur who will to take the prenatal innovation to market. At gate III the decision is to launch the product and commit the company and its reputation in the market place.

The tollgates have a double function; they operate as control valves to maintain quality and flow in the innovation funnel and also serve as a communication tool between the innovators and company management.

The screening process is part of managing the portfolio and criteria include synergy with other ideas, availability of the required resources, and the potential timing. Timing is a key factor for success. A good idea can fail because company management is not receptive to the idea at a particular time for reasons that may have nothing to do with the idea as such. It is the task of the innovation manager to be aware of these

Managing Innovation

sensitivities and to keep the idea till the time is right. Timing for market entry is another critical issue, but the entrepreneur should address this issue in stage 3.

Optimise the supply chain
An important objective for managing the innovation supply chain is to deliver a steady flow of innovations in the market place. This requires each stage of the innovation funnel to operate properly and the transfer between the stages to be efficient. But it also means that estimates have to be made of the expected development time to achieve a balanced distribution of launches over the innovation time horizon. Balanced does not mean the same number of innovations every year, but a mixture of long and short term projects and a distribution in time and domain in line with the strategic objectives.

In an efficient supply chain the capacity and capabilities are aligned with the objectives and the spheres of interest of the stages overlap. The tollgates between the stages should not lead to 'silo operation', where each stage operates in isolation and the transfer of the project takes place via documents only.

An important aspect of managing the supply chain is to monitor the position of the projects in the innovation arena. If the project crosses the boundary between inside- and outside-the-box a change in business process can be opportune.

Executing the innovation strategy

Link with company strategy
The second critical job for the innovation manager is to craft and manage the links between the innovation effort and the company strategy. Innovation is not a stand-alone activity that may or may not give valuable results. Innovation is a strategic tool that is used to achieve specific company objectives. The processes for creating the strategic objectives for innovation, for identifying and agreeing the innovation domains, and for defining the required competence portfolio need to be managed. Selecting the right innovation domains is an important issue. Many outside-the-box innovators expect to find gold in the 'white space' in the innovation arena, the area far away from the existing business domain, but it is more effective to explore

Insight in Innovation

the areas in between and just beyond the boundaries of the existing businesses and to stretch them in line with available and developing competencies. Innovation without access to the required competencies is a recipe for failure.

An innovation in isolation is always worth a lot less than one in the context of an ongoing business, and radical innovation should in the first place focus on creating opportunities by stretching the boundary of the business, finding synergy with neighbouring businesses, or creating paradigm changes inside the business as illustrated in Figure 3.2.

Identify stakeholders
Part of the strategic efforts of the innovation manager is the identification of all the possible stakeholders in the potential innovation, both inside and outside the company. Involving the stakeholders at an early stage in the innovation project and absorbing their interests in the project is an important factor for success. It is important to repeat this exercise every time the project changes direction. Involvement of stakeholders should not only be done for acquiring input and feedback, but also for looking for opportunities to involve them in the development or the financing of the innovation.

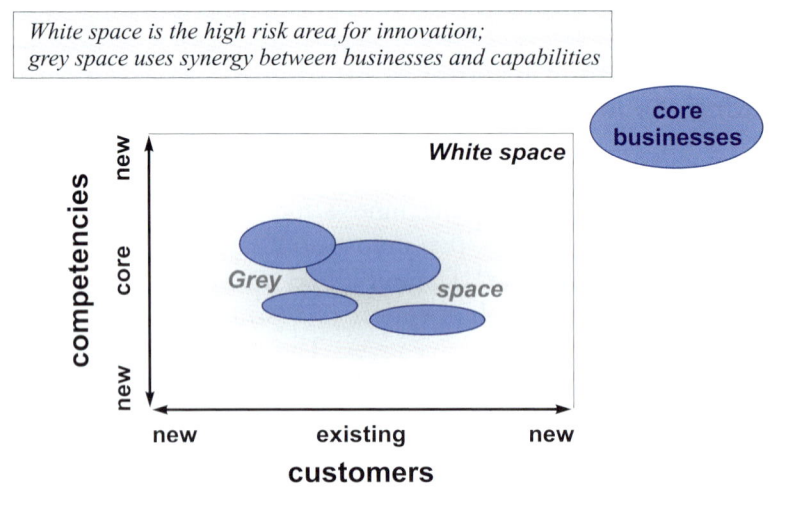

Figure 3.2 Stretching the business boundaries by exploring the grey space.

Managing Innovation

Balance the efforts
Another strategic effort is to keep the balance between top-down and bottom-up effort. As said earlier, top-down is important to maintain focus and commitment, bottom-up is important to extract creativity and create momentum. The right balance is also important to give space to the innovators and provide containment to the dynamics between convergent and divergent processes or between chaos and control.

Separate and integrate
The position of the team of out-of-the-box innovators, developers and entrepreneurs in the company is a delicate issue. On the one hand they deal mainly with ideas and projects that are new to the business and in that sense it is important that there is a degree of separation from the mainstream activities. On the other hand the innovation effort is part of the company's strategic effort and throughout the innovation funnel advice and support is required from the professionals in the business and service units. The right balance allows openness to outside ideas as well as commitment to the company's top- and bottom-line. This balance may differ from stage to stage. There is evidence that for a corporate venture unit separation is a key factor for success, at least in the first years of operation [32]. In a company where innovation is a dominant part of the company strategy and determines the company culture, separation will not be an issue, because openness to the outside world will be part of the company culture.

Managing the innovation portfolio

The third essential role of the innovation manager is to optimise the value of the innovation portfolio in relation to the innovation effort and ensure that the portfolio has the potential to meet the company objectives. The value of innovation is to create a range of options that the company may use to create competitive advantage or respond to changes in the business environment.

Assess value and relevance
A portfolio approach to innovation can help to manage the risks inherent to outside-the-box innovation. Radical innovation projects

have low probability of success and with a portfolio of options the chance that the overall effort is beneficial improves, because the successful projects can compensate for the cost of the failed projects. The basic rule is that the number of projects in the portfolio must be large compared to the failure/success ratio.

Other criteria for assessing the portfolio may vary from company to company, but a few generic criteria are:

- Does the portfolio reflect relevant emerging trends in society, business and technology
- Does the portfolio optimise the balance between business opportunities and access to competencies and capabilities
- Is the total effort in line with the required impact on the business
- Is the risk profile of the portfolio acceptable
- Is there a degree of continuity in the expected time line for market entry of the projects.

It is useful to distinguish three sub-portfolios, one for each stage in the innovation funnel: a portfolio of ideas, a portfolio of experiments and a portfolio of ventures. Each of these portfolios has to be managed in its own right. It is important that all the stages have a balanced portfolio in order to keep the momentum in the overall effort, and to make best use of available capabilities and resources. Effectively portfolio management is getting the maximum value from the available resources and competences. Available resources and competencies include those of third parties to which reliable and cost-effective access has been ensured.

A simple tool for managing the best use of capabilities is 'innovation platforms'. Innovation platforms are clusters of innovation projects with similar objectives and competence requirements. The use of innovation platforms provides focus to the overall effort in line with company strategy, limits the range of required capabilities, and supports synergy between the projects.

Assess effectiveness
There are no established techniques as yet for measuring the effectiveness of the innovation process for outside-the-box innovation. The benchmarks offered in the market place are more geared to inside-the-box innovation, typically not quantified and do not necessarily reflect best practice in innovation management. At this stage it seems

Managing Innovation

best to design a monitoring systems tailored to the specific management process. Key aspects to monitor quantitatively are:

- Are the tollgates operating effectively, i.e. stopping the bad and passing the good ideas
- Are the projects moving efficiently through the tunnel or are they stationary or recycling
- Are projects on average increasing in value through the tunnel
- Is the value-cost ratio in line with the risk profile of the innovation portfolio.

Besides quantitative monitoring, quality management checks should be carried out on the business process as such. Innovation management is a business process that should be part of the company-wide quality management system.

INNOVATION INFRASTRUCTURE

There are a number of elements that are important for the success of the innovation effort, but are beyond the direct sphere of influence of the innovation manager. These elements need to be in place to support the innovation activities and enable them to be carried out efficiently. Together these elements create the innovation infrastructure. The infrastructure is built up from many smaller and bigger elements, but we will only discuss three managerial elements: understanding and appreciating the nature and process of innovation, aligning the organisation of innovation with the company structure, and reward and recognition.

Understanding innovation

Sound innovation management is based on understanding the nature of innovation, not on following a series of prescriptions. But it is not sufficient for the innovation manager to understand innovation; all the actors in the company should understand the nature of the innovation processes and share the same appreciation of its role and purpose.

For instance, if the fundamental difference between inside- and outside-the-box innovation is not appreciated, there will be continuous pressure to amalgamate the two budgets or manage them in the same way.

Another example is the need to appreciate that innovation management is not a form of enhanced research or technology management. These processes are fundamentally different. Research management relates to managing an agreed set of exploratory or development activities that may be done on behalf of internal or external customers. Traditionally, generating and defining these activities and objectives was often left to research management; this is efficient but incestuous. Technology management is about providing access to the required technologies at lowest cost, and maximising the value of, or income from, the technology assets that the company owns.

Innovation management optimises a multi-functional business process for moving ideas to the market. Radical innovation cannot be delegated to the research or technology manager, but needs a cross-business commitment and involvement of the CEO and his management team. Delegation to the research manager may have been an option in the traditional, technology push approach to innovation where the first two stages of the innovation process were part of the research domain and inventing was seen as the first step of innovation.

Organisational principles

Obviously the innovation infrastructure is part of the company organisation and there is no point in trying to create a unique innovation structure that will be an alien element in the company. There is a correlation between a company's business position and strategy, and the type of organisation and management processes. For instance, there is little value in starting a game-changing innovation effort if the company is in consolidation mode and the main strategic objective is cutting cost. Outside-the-box innovation works better in a networking type of organisation than in a functional one. Table 3.4 illustrates the issue with a few tentative links. The links are certainly not causal and one-to-one, but based on preferences and affinities.

An important element in a good innovation infrastructure is the right approach to technology management in line with the way innovation is managed. Van Aken and Weggeman [33] provide a simple

Managing Innovation

Table 3.4 Innovation structure and business strategy

Business strategy	Consolidation	Increase value	Change the business
Innovation mode	Internal	Inside-the-box	Outside-the-box
Deliverables	Optimise operations	Develop customer value propositions	Create game-changers
Strategic driver	Cost efficiency	Company value	Growth or resilience
Organisational structure	Functional	Matrix	Network

matrix on the ways technology should be accessed depending on the technological strengths and strategic intent of the company.

Figure 3.3 shows that if the company operates in an important but unfamiliar territory the best way to access technology is via networks. The authors argue:

> *"Productivity of knowledge-based collaboration is stimulated if partners collectively share a strategic learning ambition and are able and willing to trust each other. Open informal learning alliances, i.e. innovation networks, are thus used to create options for future new business."*

This type of approach to accessing technology is an essential element for outside-the-box innovation.

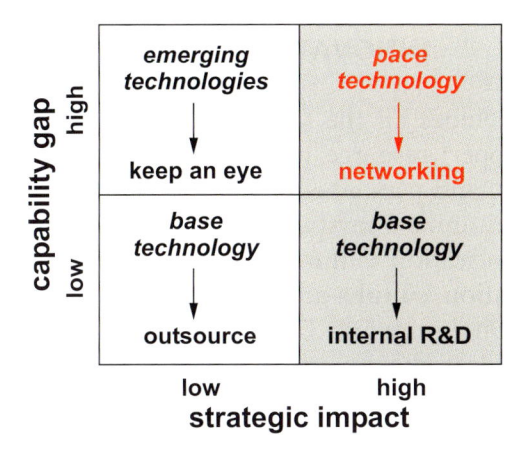

Figure 3.3 Technology development matrix.

Reward and recognition

Reward and recognition need to be part of an innovation infrastructure, but whether special rewards are necessary for innovation is a contentious issue. There is a body of opinion that argues that incentives are needed in the idea generation stage to stimulate the creation of ideas and that special compensation is required at the end of the funnel to recognise the risks that an entrepreneur has to take. The opposite argument states that each company has a wide variety of activities and innovators should not have special access to additional rewards.

A prudent approach seems that rewards for innovation should be part of the normal company reward system and not a special one. Innovation should be embedded in the company and not be treated as a special phenomenon.

However, in many companies rewards and promotion opportunities are linked dominantly to the contribution to the bottom line. But innovation is aimed at growing the top-line and the-top line should thus also feature in the company reward structure. The bottom-line of an innovation process is never in the black, and as result the rewards and career opportunities for innovators can be below par and this will have a negative impact on the whole innovation effort. The absence of recognition for innovation in the company's reward system, rather than the presence of special rewards for innovation is the more important issue.

INNOVATION CULTURE

Infrastructure includes all the facilities required to support the innovation effort proper: resource management, business processes, management systems, and personnel policies.

Innovation culture is a mixture of tangible and intangible elements. The concept 'innovation culture' is rather contentious. The issue of the right innovation culture and infrastructure has been extensively discussed in the literature [34–37]. Many sensible comments are made, but there is no clear picture emerging yet on the issue as to what is the right culture for outside- or inside-the-box innovation or on the impact of innovation culture on the effectiveness of the innovation effort.

Managing Innovation

A pragmatic approach is to assume that, if the infrastructure is adequate and the innovation effort is properly managed, the right innovation culture will result. It is often stated that:

'innovation = invention + entrepreneurship'

and in line with this equation it is suggested to equate innovation culture to:

'innovation culture = innovation infrastructure + good management.'

In other words, if the right things are in the right place and used properly, the right culture will develop or exist. This may not help much in understanding what an innovation culture is, but it avoids wasting management time on what cannot or need not be managed.

An innovation culture should include the ability to generate and manage new ways of working and new ways of thinking. A good starting point for creating this openness is to develop new ways of working by constructive exchange of know-how with external partners and by developing a willingness to understand their way of thinking. External input is an essential prerequisite for generating out-of-the-box ideas, obtaining feedback on customer needs and societal acceptability, leverage available competencies or creating access to new ones via networks, partnership and alliances. The challenge in this interactive play is to grow the know-how, without contamination or uncontrolled leakage of confidential information.

One of the most damaging elements for innovation is a blame culture. Most innovation projects fail and blame for failure could become the rule rather than the exception. Stopping an innovation project in time for the right reasons needs to be recognised as a success. Innovation is a high-risk business activity and failures will be more frequent than successes. This is a distinctive difference from all other activities in a company and as a result not always appreciated. Therefore, it is essential that risk and failure in innovation be appreciated properly, not only within the innovation community, but also throughout the company. Failures should be viewed as opportunities to learn. The seventh law of innovation is key to the right innovation culture.

Law VII Managing risks and learning from failures are key factors for success in radical innovation

This law underlines the importance of the requirement that innovation needs to be part of a continuous effort and specific strategy. Only in such a context can one learn from failure; random failure of a single project has no learning value. The distinctive features of any culture reside in the values that create and maintain the culture. Possibly the most important values to honour and protect in innovation are the '3 Ds' of **Diversity, Divergence, Differences**. While in other parts of the company consensus might be valued highly, in the innovation process it is essential to value differences of opinion and different ways of looking at the problem. The diverging stage of the innovation process wherein all the potential options are explored is an essential step in finding the best solution. Too often this activity is curtailed to cut cost or in the mistaken belief that the best solution has been found already. Innovation does not flourish in a command and control environment, but needs empowerment; it thrives on *"the discretionary energy of those who create the know-how in innovation"*.*

SUMMARY

1. Innovation is a supply chain process that needs to be managed as a specific business process.

2. The innovation process needs to be stage-gated to separate the stage-specific modes of operation, optimise the use of resources and make the high level decisions at the right time.

3. The idea generation stage has three distinctive steps: generating, capturing and nurturing ideas. Diversity and creativity are essential requirements of the ideation stage. The key objective is to identify the value to the customer of the ideas.

4. The development stage also has three stages: exploring all possible technology/market combinations (diverging step), developing the best combination (converging step) and demonstrating the

---

*G. A. Lewin, 2002, President of Shell Global Solutions.

Managing Innovation

feasibility with a working prototype. A high-level champion is core to survival in this uncertain stage. The key objective is to reduce the risk to a level that investments can be made.

5. The final stage in the innovation process is the entrepreneurial stage with three key aspects: investing in the supply chain, actively managing residual risks, and preparing detailed business and launch plans. The key objective is to successfully launch the new product with minimal risk to the company's brand and reputation.

6. The roles of the innovation manager cover three areas: supply chain management, executing the innovation strategy and optimising the innovation portfolio.

7. Efficient and effective innovation need a supporting innovation infrastructure that fits the company strategy. A good infrastructure and a well-managed innovation effort create a good innovation culture.

8. The absence of a blame culture is an important cultural element; timely stopping of a project for the right reasons should be recognised as a success.

– 4 –
Innovation and Entrepreneurship

IMPLEMENTING THE INNOVATION OPTIONS

The 'Valley of Death'

The last phase of implementing an innovation is often the hardest. The end of stage 2 was a point of triumph, with the technical and commercial feasibility demonstrated and approval obtained to enter the commercialisation phase. However, suddenly the music stops, when it dawns on the innovation team that it is not clear who is going to bring the innovative idea to market.

Effectively the only thing that has been created at the end of the development phase is an option: an option on a business opportunity. At the transition from stage 2 to 3, the option has to be 'sold' and the problem is to find a 'buyer' for that option. This point in the innovation funnel earmarks an essential difference between inside- and outside-the-box innovation.

For inside-the-box innovation the transfer at tollgate II is fairly simple from a management perspective. From the start, the project is a joint exercise between the marketing and development managers of a business unit and at this point they simply swap lead position. Simultaneously, the development team is reduced in size and the commercialisation team strengthened. The project is part of an overall marketing strategy for a product and although the plans may need to be adjusted and re-evaluated in stage 3, it is never in doubt that the marketing manager will be in charge of the final stage of the innovation process.

For outside-the-box innovation the position for the project is completely different. The question of who will be the entrepreneur is a central issue. In a perfect world this issue should have been addressed and resolved at the end of stage 2 for approval by the tollgate custodians. However, in practice this issue is usually brought to the attention of the tollgate custodians for their advice and judgement, because the issue is complex, has strategic implications and needs top-level steer.

Finding the right entrepreneur makes the transition from stage 2 to stage 3 the most intricate part of the whole innovation process and involves a step change in commitment. In the first two stages, the costs tend to be limited and are reviewed by senior managers who need to think long term, whereas in stage 3, marketing managers with a short-term focus have to take on a project with a high-risk profile. This is a difficult hurdle and the process step to find the risk-taker is sometimes referred to as the 'valley of death', because many good innovation projects get stranded here. However, the escape option of entering stage 3 without an entrepreneur and 'bypassing the valley of death' is even worse. This was often done in the classical innovation model, with research continuing the project even if marketing was not ready to take over the project. Entering stage 3 without a professional entrepreneur reduces the probability of commercial success at the same time as the financial commitments increase, whereas the chances of finding an entrepreneur do not get better. Managing innovation includes coping with the ninth law of innovation.

Law VIII Entrepreneurship is the scarcest resource in game-changing innovation

The key message from a process management perspective is that the transition at tollgate II is the critical part in the innovation process that needs strong management push and support. Passing the 'valley of death' safely needs creativity to find win-win conditions that are attractive for both the company and the entrepreneur.

Monetising the options

The first step in answering the question of *'who will bring the innovation to market'* is to look at the position of the project in the innovation arena. Obviously the project started out in the outside-the-box domain, but most projects change position during the development phase.

Innovation and Entrepreneurship

Chances are that during the development phase, the project has moved closer to the area of interest of an existing business. For instance, an early market entry in a niche of an existing market shows a better cash flow than the originally planned grand entry in a completely new market. Or in the meantime one of the business units has stretched its business boundary in recognition of the same trends in the business environment that initiated the project in the first place. The boundary between inside- and outside-the-box is a dynamic one and the distance between the position of the project and boundary needs to be assessed regularly.

The second step on the entrepreneurship issue is to list and rank the potential entrepreneurial options. Generically, there are four different approaches to entering the market and monetising the business opportunity options, and they have more or less a standard order of preference.

- Use an existing business unit
- Establish a new business unit
- Create a joint venture
- Sell the option as such

In general the most profitable way to monetise the option is to incorporate the innovation into an ongoing business, because the value of the option will be highest if it fits in a related portfolio of activities; market penetration can be earlier and faster if existing channels can be used. Also, keeping the innovation inside the company will keep the value to the company high. Selling the option, or part of it, to a third party may decrease the value by a factor of two or three, unless it fits better in the business of the buyer. If a partner that adds value to the option can be found, setting up a joint venture is a winning approach. But if the buyer is a venture capital provider or investor without a related portfolio, a large part of the value of the option will need to be transferred to create a win-win situation and to compensate for the higher risk of a stand-alone, start-up company.

Of course it could be that a number of potential monetising approaches are not available or have a different priority as a result of prior agreements with partners. It is quite possible that the agreements with partners in the development phase express the intention or commitment to continue together in the commercialisation phase. However, it would still be the right point in time for both sides to assess whether the prevailing agreements still fit best with the future of the project.

Use an existing business unit

Usually the first choice to monetise the option is to sell the innovation option in-house into an ongoing business. However, most business unit managers will react reluctantly to a proposal to bring a radical innovation to market, because the project is not part of the business plan or does not fit into the product portfolio. Radical innovations carry higher risks and slower rewards than incrementally improved products.

Most radical innovation will have an uneasy fit in existing businesses, and there may not be an obvious place in the organisation that should take responsibility for the last step.

Establish a new business unit

If none of the business units in a company is willing to take the risk, the next-best option is to make this innovation process step a corporate issue. This approach maintains the value of the option to the company, and avoids the high risk of a single, outside-the-box project for a business unit by spreading the risk over a corporate innovation portfolio. This approach involves setting up a new business unit and finding an interpreneur, the internal or in-house entrepreneur. The interpreneur can be a rather elusive figure. He is the person who can think small in a big company [69], can start a small business, and at the same time manage the big company politics. This tends to be a rare combination for an employee in a large company, since many entrepreneurs prefer to start their own company without the constraints of a big company. However, in principle, the interpreneur could also have the better of two worlds; he could combine the freedom of action of the entrepreneur in a small company with the opportunities offered by a big company, such as easy access to professional advice, marketing channels and financing support.

One of the main reasons that the interpreneur can be so elusive is the asymmetric risk-reward profile that can exist in big companies for entrepreneurial activities. Why should an experienced, successful manager become an interpreneur? The downside can easily be bigger than the upside. Most new ventures fail, even if managed properly, and this may stick to the reputation of the manager and hamper his career. Many companies reward management primarily on contribution to the bottom line and start-up ventures are notoriously poor performers in their first years with respect to the bottom line. Thus in a company that does not reward risk-taking, the interpreneur will remain elusive.

Innovation and Entrepreneurship

An innovative company needs to develop a reward structure that rewards risk taking or at least diffuses the risk of the interpreneur. One way to diffuse or resolve the risk issue is to create a corporate venture unit. This is a special corporate unit, with access to internal or external venture capital to start new business units in a professional environment, but without the entrepreneurial risks to the staff. The risks to the 'entrepreneur' are diffused by managing a portfolio of ventures and the rewards are increased because of the larger span of control and the size of the portfolio. However, the venture unit approach can take away one of the main drivers for entrepreneurial success: the commitment and drive of the entrepreneur to overcome hurdles and stick with the venture in hard times. A dip in the development of a project may send it temporarily to the bottom of the ranking order in the portfolio of the venture unit and make it a prime candidate for discontinuation and serve as the proverbial slice of the salami.

Birkinshaw, van Basten Batenburg and Murray [32] make a useful classification of corporate venture units.

- *External Financial* – Investing in external business opportunities primarily to deliver financial returns to the parent company
- *External Strategic* – Investing in external business opportunities for strategic reasons, such as windows on new technologies or to increase growth potential for the parent company
- *Internal Growth* – Investing in internal opportunities for growth, and for internal reasons such as creating a more entrepreneurial culture
- *Internal Spin out* – Investing in internal business opportunities as a means of leveraging IP and spinning out businesses that do not fit.

This classification fits very well with our classification of innovation and the four-box approach as indicated in Figure 4.1. The venture unit that fits outside-the-box innovation is the 'external strategic' variant. The unit is an integral part of the innovation funnel and is the preferred outlet for innovations that are beyond existing business boundaries.

In a purist analysis, the other types of venture units are not really innovation efforts. The two variants with financial objectives try to make money with internal inventions or with external innovations that could not be used by the existing businesses. The strategic use of

Insight in Innovation

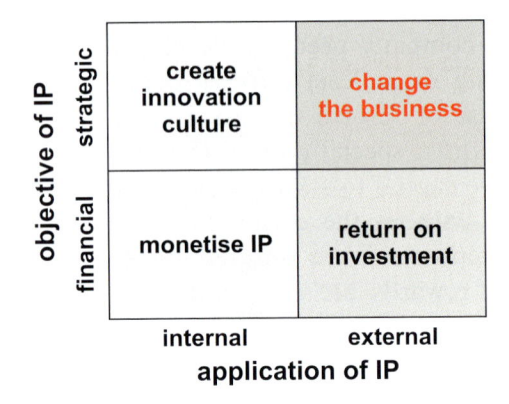

Figure 4.1 Classification of corporate venture units.

[Figure: 2×2 matrix with axes "objective of IP" (financial/strategic) and "application of IP" (internal/external). Quadrants: create innovation culture; change the business; monetise IP; return on investment.]

a venture unit to create an innovation culture is an indirect method that can have value, but typically it is an interim measure.

In Shell both the existing and the new business options have been tried and used with success. The story of the creation of Shell Hydrogen provides an illustration of the creation of a new, independent business unit under conditions where alternative options were also possible (see box on page 79).

Create a joint venture
Another approach to monetise the business opportunity option is to find venture capital and entrepreneurship outside the company. If the new partner only brings in capital, this approach reduces the value of the option, possibly by a factor of two or more, but it will limit the risk to the company and with a joint venture approach and/or buy-back option a claim on future profits can be maintained. If, however, the partner also brings in expertise and access to new markets the value of the option could increase.

This situation often occurs for small start-up companies that have a promising invention but no access to markets, and selling to or partnering with a big company is the fastest and most profitable way to monetise the option. A partnership of a start-up company with a major global company usually does not create a balanced solution, and selling tends to be the more stable and preferred option.

Partnerships that are based only on reduction of risk or provision of capital tend to have little added value to the parent company. The new partner should add value to the enterprise by bringing in

THE SHELL HYDROGEN STORY

Shell Hydrogen (SH) was created in 1998 following a so-called FRD (Fast Results Delivery) project. The findings of the FRD that the fuel cell was a disruptive technology with more chance than ever of becoming a commercial success and that hydrogen could become a major fuel in the future were fairly non-controversial. The issue was how Shell should respond to this challenge.

The first choice to make was between doing nothing and buying in when the hydrogen business proved to be successful as a low-risk, high-cost option (the wait and buy option) or participating in the development (the early adopter option) with the risk that money and management time would be wasted if the hydrogen markets did not develop. The latter option was more aligned with the overall Shell strategy of preferring active participation in development to get to know the business at first hand.

The second choice was to go it alone or do it in partnership. Partnership appeared to be a less attractive option for a number of reasons. The prime application of hydrogen is in the fuel market and the fuel market as such was too important for Shell to consider sharing. Furthermore, Shell already was a major hydrogen manufacturer for its own activities and had developed a few breakthrough hydrogen technologies.

The final choice was whether to make Shell Hydrogen part of the existing fuel business or to create an independent business unit. One of the arguments to go for a separate business was based on the belief that for disruptive technologies separate businesses are often more successful, based on the evidence given by Christensen [38].

"There is a much greater history of success for companies with disruptive innovations who have done this by spinning off the business into a separate unit, particularly if the new product may cannibalise the existing product."

Therefore a new business entity was created with the mandate to drive all Shell's global activities in hydrogen and fuel cells and to become a profitable business in the longer term.

In hindsight the decision to make Shell Hydrogen a separate business rather than a new product division in one of the existing businesses proved to be right for a number of reasons.

- All businesses in Shell had an interest one way or another in fuel cells or hydrogen and a corporate approach would create the most stable platform for business development.
- The new unit needed the right to cannibalise the existing businesses

THE SHELL HYDROGEN STORY (CONTINUED)

- A new unit gave enhanced visibility, both internally as well as externally
- The unit was small, with little hierarchy and short decision lines to support fast decision making; this proved to be a competitive advantage in a fast-moving competitive environment where creating the right alliances was key to success
- Shell Hydrogen was able to create a significant brand value.

The smallness of the unit forced SH to outsource many activities. Technology development was always outsourced to laboratories of partners. Financial, legal, personnel services were bought in. Care was taken to build the reputation of SH on performance and achievements rather than hype and PR stories.

Shell Hydrogen is active in a very broad range of fields, including technology development, venture funding, demonstration projects, and standards setting in order to build up a portfolio of opportunities. Over its lifetime of about five years it has established itself as a leading company in the hydrogen field, recognised by governments, competitors and customers as a preferred partner in business.

A few key factors to success have been:

- In innovation projects, technology and business concepts are developed hand-in-hand from the start
- Customers are involved at an early stage and where possible as partner and/or co-financer
- Provision of venture funds to start-up companies assists the development of hydrogen markets and creates a broad base for cooperation
- Infrastructure issues and standard setting are addressed as coordinated efforts with all stakeholders

Like so many start-up companies with breakthrough technologies, SH suffered from delays in technology developments, a slower than expected learning curve and consequent changes in competitive position. This required special measures to keep staff motivated. The approach adopted was to keep a balance in the mix of staff between creative exploratory people who like developing new options and can compensate for a disappointment with new ideas, and more systematic development people who keep business and technology developments on the right track.

Another factor is the special career development arrangements that ensure that staff can go back into the Shell system after three to four years in such a way that the experience in SH serves as a bonus.

Innovation and Entrepreneurship

complementary competencies or assets. This will add value to the innovation option, reduce the risk of the venture and form the basis of a stable win-win platform for cooperation.

Sell the option as such
Ultimately, if no interpreneur can be found, a solution is to sell the option as such to an interested third party. This is a last resort option, because only the lowest value of the option will be realised. Selling at this point means that only the intellectual property of the innovation can be monetised and the opportunity to gain a share of the potential margin at the customer interface is lost. In a competitive environment with alternative options available the value of the intellectual property usually is only a fraction of the potential margin.

Selling only the intellectual property as a bare patent licence is in general a low reward approach on its own. It can be an option if the technology or business opportunity has little or no synergy with the rest of the business activities and does not fit the company strategy. A once-off licensing deal is not very attractive and licensing intellectual property needs to be part of a wider know-how and service package or treated as an asset as part of a technology swap or input to create a strategic alliance to be valuable.

Large companies can use a special licensing or venture unit (the internal, financial variant in Figure 4.1) to monetise these options. Such a unit can be very rewarding when the parent company possesses a large stock of unused or under-utilised intellectual property. Monetising this intellectual capital is equivalent to 'innovation with a delay', most likely the outcome is completely different than the original intention, but it can create much value for a company.

Licensing the intellectual property of an innovation can be attractive under certain circumstances. The best and most profitable condition is when the intellectual property protects a unique feature or capability of a product that has a high value. Licensing from a monopoly position is very valuable because, with a limited effort, it can claim a large share of the margins in the market.

In mature markets, monopoly positions are rare and licensing takes place in a competitive environment. Licensing margins tend to be low under such circumstances, but it can still be an attractive opportunity when the markets that can be entered via licensing are an order of magnitude higher than would be the case otherwise. This can be the case for so-called 'basic technologies' that can be applied in a wide variety of applications, for example chips in IT applications, or for small companies with low cost structures selling to big companies. In both instances the multiplier effect can be very large and more than compensates for the low value per licensed unit.

THE FINAL HURDLES

A key task of the entrepreneur in stage 3 is to manage the risks of the market entry with respect to how, when and where. In stage 2 the development work should have reduced the risks to the point that they cannot be significantly reduced further with more tests and analyses, and investments can be made with acceptable risks and meeting the usual return on capital criteria. Whereas in stage 2 the risks are mainly technological and can be reduced or eliminated by more information derived from further tests or studies, in stage 3 certain business risks in terms of launch, brand, exposure, cannot be eliminated completely and the residual risks have to be managed. Certain risks can be reduced by training and building up sales and marketing capabilities, but for those risks that cannot be eliminated an action plan on how to manage them has to be developed.

The main risks in stage 3 are generic and intrinsic to any product launch project and have to be managed by the project manager according to management techniques adopted by the company as best practice. However, there are a number of risk factors that are specific to outside-the-box innovation that need to be addressed as well. We will call them 'the final hurdles'. Finding the right time and right place to enter the market are the most important ones, but existing infrastructures, and the fight back of the old technologies are others.

Timing

In many innovation handbooks a short time to market is mentioned as the most important issue, and time to market certainly is critical, but sometimes the right time is more important than the fastest time, and keeping the invention on the shelf for a period can make the difference between success and failure.

As for any option, an innovation option should be exercised at the right time to maximise its value. This can be a delicate task, because the window of opportunity to enter the market may be limited. The earliest date is usually, but not necessarily, the best time. This is particularly true for outside-the-box innovations that are based on disruptive technologies. Many innovations that were introduced too early did not survive their premature birth. For instance, several innovations based on the Internet that were very early in the game did not make it, because the use of the Internet was not yet sufficiently widespread. Criteria for the right time are hard to give, but a good look at the environment in which the disruptive innovation has to flourish is a necessity to check that the new innovation has enough linkage points with its environment. If the innovation is very alien, it may not survive or may need a long introduction time. An outside-the-box innovation needs to fit in the environment where it will be used.

Innovation has two important cyclical elements, one as part of an internal cycle and one related to an external cycle.

The performance and profitability of an industry and companies go through good times and bad times. In hard times strategic emphasis in most companies tends to shift to reducing costs rather than stretching the business and outside-the-box innovation suffers as a result. Good ideas created in bad times have an additional challenge in the quest to the market and it may be better to lay low until better times return.

Also innovations tend to come in clusters and waves. The waves are created by the emergence of new technologies that allow the creation of new applications and often lead to increased economic activity. In return the economic activity leads to more innovations. Many futurologists believe that the beginning of the 21st century may again see an outburst of innovations based on new technologies such as life sciences, nano-technology, micro-energy, and information and communication technologies.

Often innovations are made possible by combining two or more breakthrough technologies. For instance, big aircraft needed strong, lightweight materials as well as compact, high-power engines. Many IT-related innovations need enough users of a certain communication infrastructure before they can become attractive. This type of linkage serves as a 'tollgate in time' for prospective innovations.

Getting the timing right is difficult not only for external reasons. The main internal reason for missing the right time is the management process for making the big decisions for investments and launch. These decisions are particularly difficult for game-changing innovation and can lead to delays or postponement.

The main 'reason to avoid' making the big decision is almost intrinsic to the outside-the-box innovation process. In theory the whole system is geared to finding and developing 'the big one', the major game-changing innovation that will make the company a leader in a new business with defendable high margins and a sustainable competitive advantage. In practice the innovation system often shies away from the really big innovations and ends up supporting a series of smaller ones. The really big, game-changing innovation has a high-risk profile and potentially a big impact on the company. It will probably cannibalise the existing business drastically and can have a big negative impact on the reputation and brand if the innovation is not successful. Thus the temptation will be to downgrade the really big ideas, particularly in stage 3 where the money at stake and the potential exposure to the company are biggest. The potential damage to the brand is another major reasons why big ideas have a hard time in large companies. In most aspects—competencies, finances, channels to customer, distribution facilities—large companies provide a good environment for developing and launching big ideas, and one that is better than in small companies. But one disadvantage, the large downside risk, often outweighs all the advantages. This hurdle, together with the different risk-reward profile, goes a long way towards explaining why outside-the-box innovation tends to be done via small start-up companies that have well-defined downside risks, and why in contrast large companies tend to be best at incremental innovation where they can maximise their scale advantage.

This internal conflict of simultaneously chasing the big innovations and avoiding them operates throughout the innovation funnel. Gaynor [11] gives a hypothetical, but poignant example of how brave an innovator in 3M would have to be, to push his idea to replace the Post-it Notes with a game-changing technology that would outperform the Post-it Notes by a big margin. Post-it Notes are the biggest innovation by 3M in the last decades and create very significant revenues, brand value and recognition for 3M as an innovative company. The chances to get much support to start work to cannibalise that product on the basis of a rough idea will be low and the risk that such an initiative may hamper the esteem of the brave innovator are real. Many would-be innovators faced with that dilemma will decide to go on with their normal work.

Innovation and Entrepreneurship

Infrastructures

Infrastructures and standards are important elements in innovation, either as a blocker or as a platform. Existing standards and infrastructures can be major hurdles against breakthrough innovation, because they will need to be changed or replaced and that is expensive and time consuming. Also, the existing infrastructure supports and protects the position of established products, which are challenged by the disruptive technology and these products will actively fight against changes in the infrastructure. Thus disruptive technologies often have to be introduced within the constraints of the old standards and be compatible with the existing infrastructure until they have developed sufficient momentum to support their own, new infrastructure.

The simple example that a new type of train needs to fit the existing rail system, unless the train is so good and unique that it pays to build a new rail system, is an obvious one. But the infrastructure and standards hurdle goes deeper. A new type of bottle needs to fit the crates that need to fit the pallets that need to fit the containers, and so on. New software needs to be compatible with old software, new and old connecting systems, etc. Almost all things we use are embedded in standards and supply chains that have been created to increase reliability, safety and efficiency; they provide a powerful system for continuous improvement, but are blockers for a breakthrough change.

However, once an infrastructure has been changed or a new one established, it creates a very efficient new platform for many new applications. Creating and setting new standards is an important element of game-changing innovation and this part of the process can create much value. By its nature it is difficult and usually impossible to own a standard and make money with it in the same way as with IP but establishing the right standard can be equally valuable. Figure 4.2 gives a series of examples of the development of well-known infrastructures that became engines for economic growth. Typical time to mature a major infrastructure is about a century. The beginning of the S-curve is the most fertile period for innovation; gradually innovations will change from outside to inside the box and the top part of the S-curve tends to be a barrier for outside-the-box innovation. Certain innovations will thus have a better chance in certain periods than in others. This S-curve effect not only features externally in the customer domain, it can also play a role

Insight in Innovation

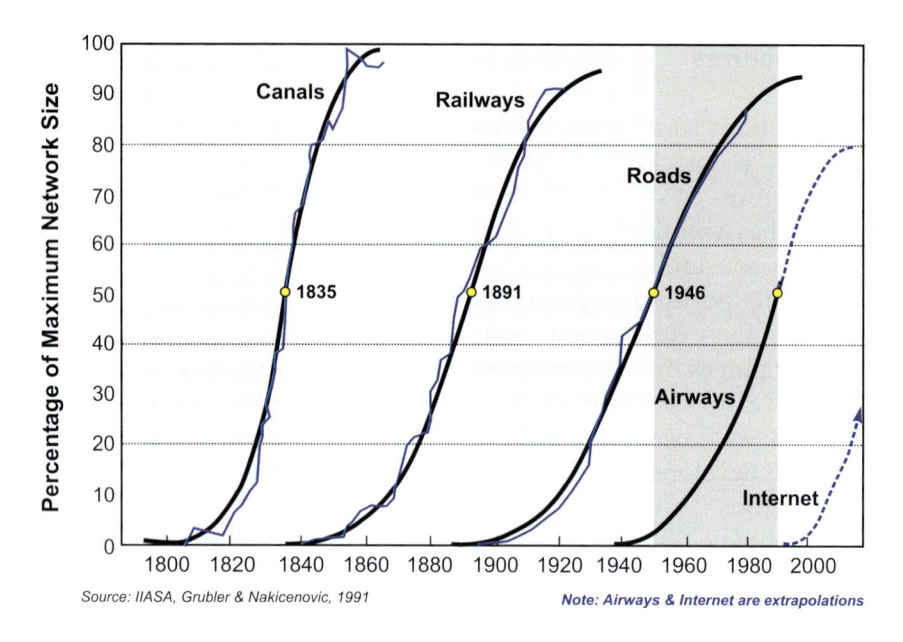

Figure 4.2 Development times of infrastructures.

inside a company or the industry. Companies and industries also move through an S-curve in time and here the bottom part of the curve is the more fertile part for outside-the-box and the top part for inside-the-box innovation.

Markets where the old infrastructures and standards do not exist have a major advantage for breakthrough innovation. Competitive advantage can be created by the simpler introduction into the market and by avoiding the cost of removing the old infrastructure and thereby leapfrogging to a new and better infrastructure, which in turn makes the competitive advantage sustainable. This type of opportunity often exists in developing countries and the rapid emergence of the mobile phone in the Far East nicely illustrates this point.

The fight back of the old technologies

The challenge that breakthrough innovation brings often inspires the existing technologies to improve significantly. And since they are in control of the channels to the market and already have an infrastructure and economy of scale, they are in an excellent position to withstand the attack of the newcomers. Thus the new technologies must have a very significant advantage over the old ones. A step change in cost reduction is important, but pure economics are seldom sufficient on their own. Moreover disruptive technologies usually have to start with a cost disadvantage. The new technology must have a performance or emotional advantage to be successful.

The fight back of the old technologies can now be seen in the rapid developments in efficiency and cleanliness of the internal combustion engine (ICE). When the development of the fuel cell car was started at the end of the last century, it promised to give a major advantage in terms of emissions over the ICE car. This advantage has been rapidly eroded and the fuel cell car will have to rely on other benefits, such as the all-electric, noiseless car, to win the battle.

But even if the new technology has a cost and performance advantage, the old technology often has the emotional advantage, not only for the customers but also for the decision makers within the company. The fight between sail and steam provides a beautiful example of this emotive aspect, where for a long time the sail ship could compensate for the disadvantage in reliability and economics by the strength of the emotional attachment to the beauty of a sail-ship in full sails on the open sea.

The battle of 'sail versus steam' also provides a beautiful example of the other wisdom from history: *the old technology dies at its peak*. The last sailing ships were better, faster and more beautiful than ever, before they were pushed out of the market by the noisy and dirty steamships.

The related consequence is that the harder the old technologies fight back, the bigger the fall when they finally lose the battle. Companies that have put all the effort on defending and perfecting the old technologies will not survive when the war is lost and the resulting consolidation process serves to strengthen the position of the newcomers.

Emotions

Emotions play an important role in innovation. The momentum to push an idea forward through the innovation funnel, to overcome internal hurdles and to survive the occasional crisis comes largely from the emotional energy of the persons who attach a dream to the idea. Innovation is at its best when it is based on a vision, and

Insight in Innovation

a vision is never built on facts alone, but requires an emotive element to glue the pieces together.

Also resistance to breakthrough innovation is more often based on emotions than on economics. Game-changing innovation is often a threat to 'the existing' and attachment to the existing becomes an emotion. This emotional element is much less of an issue for inside-the-box innovation; incremental innovation brings improvement to the existing product and improvement leads to emotional support rather than resistance.

The emotional aspect of acceptance of an innovation by customers and society is a very important element. It is the key factor for the intrinsic unpredictability of success for game-changing innovation. One of the most striking examples from history, as summarised in Figure 4.3, is the fight between the steam-driven car against the electric and the ICE cars.

Objectively the electric car was the best performer and was winning all the races. However, it missed the VROOOOM factor that gave that emotional feeling of power and speed that the customers at the time valued most. Hopefully for the fuel cell car this emotion has changed in the mind of the customer over the last hundred years.

Outside-the-box innovation often needs time to find the right fit between its intrinsic capabilities and external features. Most game-changing innovations enter the market in a conventional disguise. For instance, the first cars were converted or mimicked horse carriages. This is often the result of two factors: the novel application needs

Features	Steam	Electric	ICE
Technology	Familiar	Simple	New
Driving quality	Reliable engine, heavy, easily stuck	Quiet, clean, easy start	Noisy, polluting, difficult to start, unreliable
Fuel supply	Hardware stores	Slow recharge	Few stations
Range	High	Low	High
Emotional element		Winner of most races	The Vroooom factor

Figure 4.3 The winner is unpredictable.

Innovation and Entrepreneurship

to fit in the existing infrastructure and the customer must be able to recognise the new tool and relate it to things he knows. It takes time before the intrinsic features and advantages of the new materials and technologies can be used in the design to optimise the performance without running the risk of alienating the customers. The customer likes to recognise certain old features, which for a long time have been sold as valuable symbols of performance and status.

The emotive factors are key in the eventual success of a game-changing innovation. Innovative ideas that are recognised as big game changers from the start may never make it due to the resistance they create, whereas many of the successful game-changing innovations were introduced for other purposes or as a sideliner. The Internet was introduced to support scientific interest groups, grew for other reasons to become a generic, global communication tool; the PC was not recognised from the start as the game changer it became.

The lessons from this for managers involved in the innovation process are many:

- Be cautious in your judgement; the chances that you will recognise the big winner from the start are modest
- Involve all stakeholders from the start and improve the odds that you will recognise potential losers a bit better and earlier
- Do not manage the dream out of the idea in the innovation process; simplistic economics and strictly rational arguments will make a niche opportunity from a potential game changer
- Recognise that each radical innovation process also is a change process and learn from the change manager how to handle the emotive part of big innovations
- The market is the ultimate arbiter for success; if you want a big winner, be ready to test many opportunities and accept many failures.

Although it can be managed to a limited extent only, it is important that the emotive part of innovation is recognised in the management process. The basic equations in innovation can be adjusted to:

$$\textbf{innovation} = \textbf{invention} + \textbf{entrepreneurship} + \textbf{vision}$$

and

$$\textbf{innovation culture} = \textbf{innovation infrastructure} + \textbf{insightful management} + \textbf{vision}$$

Management of the innovation funnel can only manage the 99% perspiration part, but should recognise the importance of the 1% inspiration. Weeding out the bad ideas rationally and rigorously needs to be balanced by intuition and vision that allows potential 'possibles' to prove themselves in the market place.

SUMMARY

1. Entrepreneurship is a scarce resource in the innovation process, but do not try to pass 'the valley of death' without it.

2. The interpreneur, or internal entrepreneur, is a key resource in innovation, but he/she can be an elusive figure if the risk/reward structure is unattractive in a company.

3. Partnerships in innovation should be entertained primarily if the partner can bring additional value to the innovation with complementary capabilities or customer knowledge, and not only for reducing capital commitments.

4. The final hurdles to a successful innovation are external factors related to emotional responses by management, customers and society. These factors can be managed only to a limited extent and the most important weapon for success is 'the right timing'. You need to be in tune with the times to get the timing right.

5. Breakthrough innovations can see their initial advantage erode over time by the fight back of the old technologies. However, the stronger the resistance, the bigger the eventual victory can be, because many companies with the old technology will die in the process.

6. The market is the ultimate arbiter of success.

– 5 –
The Value of Innovation

INNOVATION CREATES OPTIONS

The innovation process creates new business options that may or may not be used in the future. These options have a value, however their value is usually not quantified and does not appear in the company accounts. Still the stock markets value innovation highly and innovative companies tend to have a premium in their market value. This premium value is given because innovation improves the resilience and long-term expectations of a company, as expressed by the ninth law of innovation.

Law IX The value of innovation is in the creation of new options for the company and new choices for the customers

The new options made available to the business by the innovation process can be used to create new choices for the customers or not used at all. The extra value of an option is that it does not have to be used and there is no penalty for not using the option. When it turns out that the option may not create a value proposition for the customer, the option is simply not exercised.

The customer is a crucial partner in innovation, but also too often ignored or undervalued in the innovation process. The statement at the end of the previous chapter 'the market is the ultimate arbiter of success in innovation' only acquires meaning if it is recognised that it is the customer who pays for the innovation and thus determines its value. The upper boundary of the value of an innovation is given by the ability and willingness of the potential customers to pay and the value that each customer can give to the innovation. That value will depend on the size of the problem or the unfulfilled need addressed by the innovation.

At the current state of the art, assessing the potential value of outside-the-box innovation and monitoring the value of the innovation project through the development stages is one of the weaker parts in innovation management. However, the old management truism *'what isn't measured, isn't managed'* also holds for innovation.

The three value domains

In order to assess the value of innovation, we have to understand the value creation process and identify the factors that determine the value. The value of innovation is determined in three different domains: the customer domain, the strategy domain and the innovation domain. These domains represent three different mechanisms that determine or limit the value of an innovation project: the customer's willingness to pay, the company's strategy on how to extract value from the market, and the innovative idea on how to create new value. The value of innovation needs to be assessed independently from these three different perspectives. In the innovation domain the potential value of the innovation project is estimated, whereas in the strategy and customer domains only the limits to the value of the innovation can be assessed.

The innovation domain. Innovation only acquires real value when it is implemented in the market. But before the launch the innovation project has already created potential value, for instance by creating intellectual property. The value of an innovation project during the development phase is derived from the potential business value in the market and should be equal to the fair market value of the innovation effort at a particular time. This figure will in general grow during the innovation process, to reach the initial business value at the launch.

The customer domain. The estimate of the value in this domain assesses the needs and purchasing power of the customer in the market segment under consideration. This estimate is based on the value that the customer has embedded in a particular market segment and provides a ceiling figure to the innovation.

The strategy domain. The estimate in this domain should establish the value of the strategy adopted by the company on how it will extract the value from the customer domain. This estimate provides another limit to the value of the innovation and should be lower than the estimated value in the customer domain.

The Value of Innovation

The potential value of the innovation should be lower than that of the other estimates, unless the innovation genuinely creates new, previously unidentified customer demand that will replace other demands and was not captured in the other estimates.

For major innovations the strategy domain and the innovation domain will be more or less the same, and so will be the value estimates. But it can be that several innovations are part of the strategy and then the sum of the value of the innovations should be equal to or less than the value of the strategy.

The link between the strategy and innovation estimates is very important because it establishes both to company management and to the innovation team the ambition level of the innovation effort, and the strategy estimate can be used as a target value for the innovation projects.

The option value provides the real value estimate for the innovation, but the two other estimates are important reference figures that provide a reality check. The customer ultimately determines the real value of innovation, the strategy sets the target value for the innovation and each project creates an option value.

In assessing the value of innovation it is assumed that the three domains fully overlap, as indicated in Figure 5.1a, and that the full value of the option can be realised in the market place. However, it is not uncommon that the three domains do not fully overlap, as illustrated in Figure 5.1b, and that only part of the potential value of the innovation can be captured. This can occur, for example, when the company strategy does not fully reach the intended market segment or the innovation projects are not fully aligned with the company strategy.

The need for full alignment between the three domains sounds simple and obvious, but alignment does not necessarily occur. The link between customer and strategy is often weak and poorly understood in outside-the-box innovation. It can also happen that the estimates for the customer or strategy domains are lower than the potential value of the innovation. This not only means that one or more of the estimates may be wrong, but also that the innovation is not properly positioned yet. Misalignment between the value of the innovation and the value to the customers can easily occur in markets where government taxes and subsidies are an important part of the price of the product. The link between price and value is distorted and this can lead to confusion. A well-known example is the market for green energy. Many forms of alternative energy look very attractive when the costs are compared to the price the customer pays, but they are still very uncompetitive when compared to the cost of fossil fuels.

Insight in Innovation

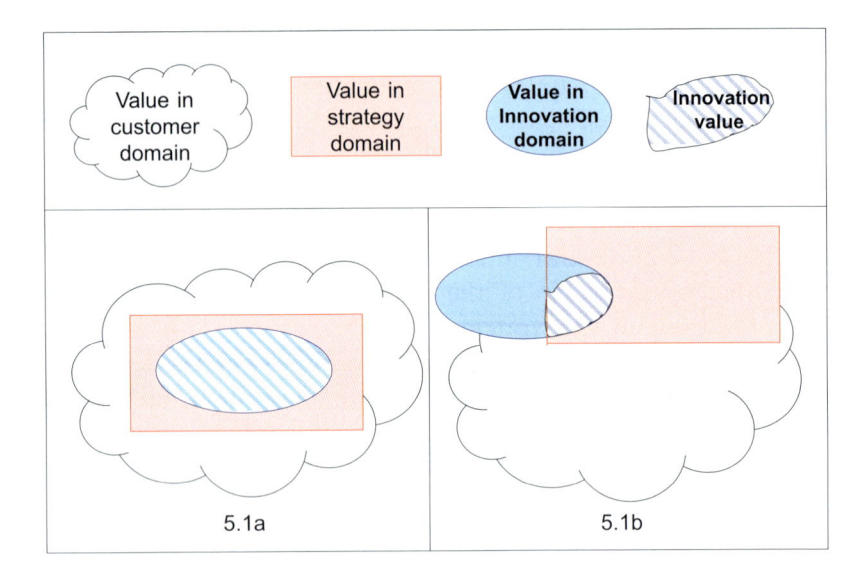

Figure 5.1 Alignment and mismatch between the three value domains.

To maximise the value of the exercise, it is important that different parties make the three estimates independently and preferably that an outside party with more intimate knowledge of that domain does the estimate of the value in the customer domain.

The separate and independent assessment of the three value domains is less important for inside-the-box innovation. Since the innovation will be a modification of an existing product and the modification is part of a product strategy, the chance of misalignment between the three domains is limited as is the probability that the value of the innovation is assessed at a higher level than the value in the market.

The roles of and the linkages between the customer, the company strategy and the innovation projects in setting the value for outside the box innovation are pointedly illustrated in the story of '*Murex and the red tins*'. This story of how in 1892 Marcus Samuel* opened

---

*Marcus Samuel, together with his brother Sam, ran a London-based trading business (M. Samuel & Co.—established in 1834). The business traded a wide range of products in the Far East, but particularly in Japan. Products included textiles, tools, machinery, rice, silk, case oil and shells.

In 1897 the company was renamed the Shell Transport and Trading Company and merged in 1906 with the Royal Dutch Petroleum Company to create the Royal Dutch/Shell Group of companies.

The Value of Innovation

MUREX AND THE RED TINS

Under cover of a routine business trip to his trading houses in the Far East, in 1890, Marcus Samuel secretly travelled to the Baku oil fields in Russia to explore the potential of exporting kerosene to the growing markets in the Far East. At that time Standard Oil controlled about 80% of the global kerosene trade and defended its monopoly position ruthlessly with a combination of efficient, integrated supply chains, economy of scale, and focused price wars. Marcus designed a bold strategy for developing his own kerosene trading business in the Far East with a defendable competitive advantage. To break into the markets he needed to be able to do business and undersell Standard Oil throughout the region from the start. In order to create this position he needed both a reliable, high volume supply of product, low transport cost as well as simultaneous roll-out in the whole region, otherwise Standard Oil would drop prices in one market financed by increased prices in other markets. To minimise the risks the market entry strategy also required full cooperation and co-financing by his suppliers and trading partners.

The strategy was executed flawlessly. In 1891 Marcus Samuel signed contracts in the UK for the delivery of a fleet of modern ships and in Russia for the supply of oil from the Baku fields. The design of the kerosene carriers was based on breakthrough technology. Standard Oil supplied the kerosene (or case-oil) to the Far East in blue cases. Samuel wanted bulk tankers to reduce transport cost, but the bulk tankers used for crude oil were quite unsafe and only allowed in a few places. The design of the bulk tankers was improved with so many safety features to contain the risk of fires and explosions that the ships were granted permission to sail through the Suez Canal, cutting the trip to the markets by 4000 miles. In order to further reduce cost, the tankers could be steam cleaned and filled with goods on the return trip. To reduce the risks and limit the financial exposure the Russian oil was delivered on favourable credit terms and the investments in the local storage facilities were made by the trading houses.

In 1892, **Murex**, the first ship of the fleet, was launched in the UK and made its maiden trip to the Far East, picking up kerosene in Russia. By the end of 1893, Samuel had launched ten more ships, the trading houses had build an extensive storage and distribution network throughout the Far East, and the new oil was everywhere.

However, one minor oversight, the customer, almost destroyed the venture. The sales concept adopted was that the customers would fill up their old blue tins with cheap kerosene supplied in bulk. But the tins were valuable to the customers as raw material for a myriad of domestic tools such as roofing, birdcages, hibachis, tea strainers, and this value

MUREX AND THE RED TINS (CONTINUED)

outweighed the lower cost of the kerosene. Once this value of the tins to the 'domestic economy' was realised, Samuel sent a ship with tinplates to the Far East and ordered the trading houses to start making new tins, painted red. The local manufacture delivered undamaged tins to the customers, provided local employment and created goodwill for the company. The easily identifiable bright red colour created a competitive advantage, and soon the bright and shiny red tins were replacing the old and rusty blue tins that were battered by a long sea voyage.

---

the market for Russian kerosene in the Far East is probably one of the most important innovations in the history of Shell. The story of the 'Marcus Coup' has been told before [39,40], but we will tell the innovator's tale.

The story shows the different contributions of the customer, the company strategy and the innovation projects. A brilliant business strategy almost failed because the customer response was different than expected. The strategy was supply- rather than demand-driven, but it was not the strategy, but the customer who determined the fate of the innovation. In the end the red tins made the difference between success and failure, not the low-cost supply of the product or the breakthrough technology used in the bulk carrier. Murex was key to the supply low-cost kerosene to the Far East, the bright red tins were key to the demand for the kerosene. Cost reduction alone did not move many customers, but in combination with a value factor it was a great success.

Samuels' strategy supported several innovation platforms with a wide range of projects. The key innovation platforms were:

- the new business model and supply chain
- the breakthrough design of the ships, with the first-class safety rating by Lloyds and the permission to pass through the Suez Canal
- the distribution and storage facilities throughout the region
- the financing model by extended credits from suppliers and traders
- the local manufacturing of the bright red tins.

The Value of Innovation

This comprehensive array of innovations and implementation strategies could be considered as one of the founding innovations of Shell.* In his book on the history of Shell, Howarth [40] writes:

> "The lesson was clear. To outdo competition, the product must be at least as good, cheaper if possible, and somehow must also provide added value to the customer. Learned very early on, it was a lesson which, in the century since, Shell has almost always remembered, or forgotten at is peril."

The two-stage innovation of 'Murex and the red tins' shows all the value elements of a sound, out of the box innovation.

- Marcus Samuel had deep insight into the value chain and he possessed outstanding inventive and entrepreneurial capabilities
- The innovation is based on a 'new combination' between a new capability – safe bulk transport of kerosene through the Suez Canal – and a new market (for the company) – kerosene in the Far East –
- The time to market is very low, only 2 years
- The innovation is strategy driven
- A new business model is the value driver for the innovation; breakthrough technology is an enabler, not the driver
- Risks are actively managed with creative financing methods and cooperation with partners
- Suppliers and partners participate in the innovation process
- The customer was involved, albeit a bit late, and the innovation was provided with an emotional factor as well as cost reduction.

The story of Murex and the red tins shows that the overlap between the three 'value-creating' domains was not perfect in this case. Marcus Samuel assumed that the market for bulk kerosene and case oil was more or less the same, but this was not the case and the lack of overlap reduced the value of the 'Murex innovation'

---

*Great innovations can determine the culture of a company for a long time. Rockefeller innovated the vertically integrated petroleum company and economy of scale to create dominant competitive advantage. His achievement was described by one of his successors:

> "He instinctively realised that orderliness would only proceed from a centralized control of large aggregations of plant and capital, with the aim of an orderly flow of products from the producer to the customer."

This cultural heritage still brings value to Rockefeller's descendants.

considerably. On the other hand, the market for oil in bright cans was bigger than the original case oil market and the value of Murex could be maximised.

The story also shows that the two ceilings to the value of an innovation are determined by complementary mechanisms.

- *The 'Murex' side* (*or supply or company strategy side*): The value of the innovation strategy is based on how much product the company can supply to the market. In this case the value was related to the capacity of the fleet of bulk carriers and the volume the traders in the Far East could handle.
- *The 'red tins' side* (*or demand or customer side*): The value of the innovation is based on how much product the market can absorb and pay for. In this case the value was given by the number of tins that the customers were willing or able to buy.

Furthermore, the story also points at an important value element of outside-the-box innovation that is not always recognised. All outside-the-box innovations are potential platforms for further innovation; the new application in the market will lead to new opportunities and new needs. For example, the Murex innovation led to the red tins and although at first seen as a necessary extension of the supply chain for the Russian kerosene, the bright red tins soon created their own value proposition and became a platform for further innovations. This platform effect can make the difference between a marginal and a robust project.

Finally, the story underlines that the line between success and failure is very thin. If the value of the tin to the customer had not been recognised in time or if the response had been inadequate, business history would have been different.

ASSESSING THE VALUE OF INNOVATION

The purpose of an innovation project is in creating an option that allows a company and ultimately the customer to make a choice. Thus the best way to measure the value of an innovation before it is launched is to estimate its option value. The option value reflects the business value of an idea, discounted for the chances of success, the time required to bring it to the market, the uncertainties in the development path and the market conditions, and the decisions that

The Value of Innovation

management can make in the future. Typically, customer and application are not clearly defined at the start of the project, but will shift during the process, sometimes drastically, and thus the option value needs to be estimated regularly. A regular assessment of the option value allows players to track the progress in how much value is added to the idea over time and facilitates timely stopping of a project. The options increase in value if the likelihood of success is estimated higher, reaching its maximum value when the option is exercised and leaves the portfolio. They decrease in value when the chances for success diminish and when the option is ultimately rejected the development costs have to be written off, but otherwise there is no other penalty for not using the option.

Many people are reluctant to use quantified values to measure the impact of innovation, because the data used have such a large degree of uncertainty and often a built-in optimism on margins and profits that never materialise. Quantifying the value of an innovation project is not simple, considering the uncertainties involved, but it is certainly possible. It is important that the value of innovation is expressed in the same currency, i.e. money, as used for investments in the rest of the business; it is the most effective way to position innovation properly on the mental map of business management. Quantification of the value of the innovation effort is also good management hygiene, because it ensures that the key stakeholders are identified and agree with the anticipated costs and benefits at an early stage. As with many quantification efforts the value of the joint exercise as such is often as important as the result.

The value of an option

The simple way to value an innovation project by calculating a NPV (Net Present Value) with a special factor for the risk element can be adequate for inside-the-box innovation, but does not do justice to the value of an outside-the-box innovation effort. The NPV method calculates the value of a predetermined development path, whereas the option value approach includes the decisions management can make during the process in order to reduce cost or risk; for instance, to stop the project when the research results or market expectations are disappointing before major investments are made, or to change the direction of the project [41].

Insight in Innovation

For example, Marcus Samuel could have estimated the value of the kerosene bulk carriers at the design stage with a simple NPV calculation using a high discount factor for the high overall risk that the whole venture would fail for one reason or another. Alternatively, he could have calculated an option value for the shipping venture using specific risks that could occur at specific stages in the project including the decision not to build the ships and stop the project if the design did not get the safety certificate required to pass through the Suez Canal. This opportunity to avoid a major unprofitable investment increases the value of the innovation project and illustrates why an option value is a more relevant figure than a NPV figure. The value of options resides in waiting for better information before decisions are made [42].

Calculating a proper option value with a comprehensive financial model is not easy due to the many uncertainties and the resulting complexity. It is possible to use simpler stochastic models to obtain an approximate option value that still reflects the various decision points in the development process, as well as the key risks and uncertainties.

A relatively simple approach is to map a number of possible developments, quantify the possible outcomes and assess the risk of success associated with each decision point and the uncertainties in the possible outcomes. Figure 5.2 illustrates this approach. The overall

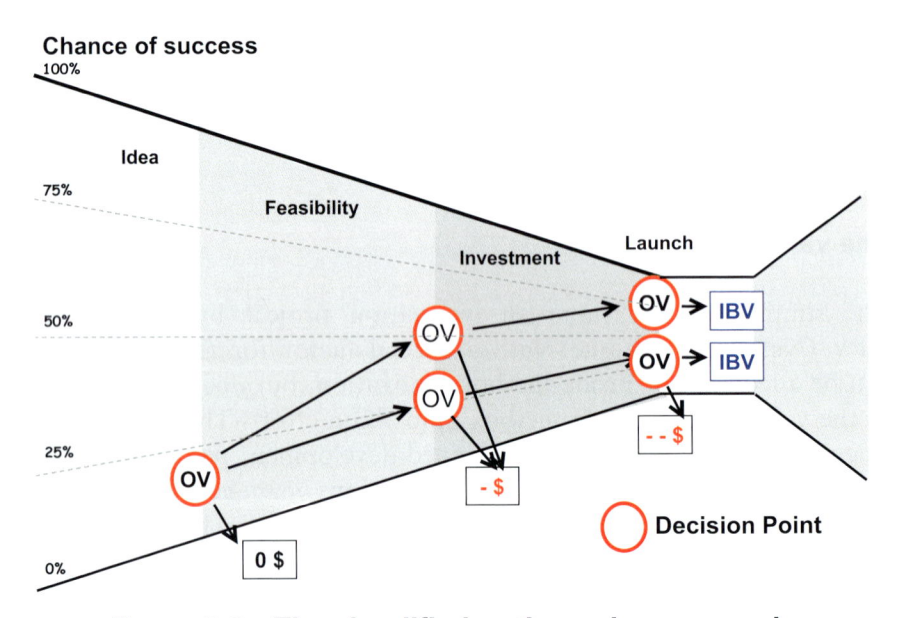

Figure 5.2 The simplified option value approach.

The Value of Innovation

outcome can then be calculated with a stochastic model. Uncertainties that need to be reflected in the model include the growth rate, margins, market share and their changes over time, the chance of success, the timeline of the project, etc. An additional advantage of the simplified option value approach is that it can be evaluated only on net added value in the segments of the supply chain and in this way the large uncertainty in the cost of the required investments in physical facilities can be avoided in the early stages of the innovation process.

The value of the option will vary during the innovation process. In a perfect world, i.e. a development process without directional changes of the development or other surprises, it should increase gradually from a notional value at the idea stage to the business value at the time of the market launch, mainly as a result of the reduction in risk. In practice the value of the option will be very volatile in time, because development costs and perceptions on potential margins will change.

The process for estimating the option value has four main steps:

1. Identify key income parameters and future uncertainties
2. Quantify the income and uncertainties
3. Identify the key decision points
4. Run the option model and analyse the results.

As with many assessments the value of the actual quantification is always limited because the input data have limited reliability. But the value of the quantification is always worth the effort, as it still represents the best estimate of the value of the innovation effort and money is the best language for comparing value.

It is not suggested that the option value of an innovation project be used as the only parameter to decide whether to continue the project. A positive option value indicates only that statistically the innovation project is expected to have rewards that are higher than the costs. Obviously, a high value will be supportive of continuing the project and a low value a consideration to stop, but the volatility is high and the spot value on its own is not very reliable. The trend in the option value can be an important parameter in combination with other considerations such as strategic fit, potential revenues or platform for other innovations.

Option values are always low compared to the size of the prize. If outside-the-box innovation is measured only in contribution to the bottom line, its value may be underestimated and suffer from the 'innovator's dilemma'. This dilemma has been eloquently expressed by Christensen [38,43] who argued that the value of disruptive technologies tends not to be recognised by analyses and criteria that are used successfully for inside-the-box innovation. The reason is that new products based on disruptive technologies usually start out with a poorer performance-cost ratio than existing products, thus penetration into existing markets has a low chance of success. Knowledge of alternative markets or factors via which the new products could grow and improve is limited and the probability that the new product could enter other markets at a later stage via the backdoor is underestimated.

Another way of explaining this dilemma is that because the developments of both the old and new product follow an S-curve in time and the S-curve of the new product starts at a lower level than the existing products, the prediction of the take-off point of the new product and the intersection with the old product is difficult with the limited data available. Thus it is important to also measure the value of the innovation by the impact on the 'top line in the future' and not only on the 'bottom line now'. By keeping an eye on the size of the prize, i.e. the potential contribution to the cash flow in the future, a more balanced platform for judgement is created.

The difference between the option value and the size of the prize analysis is that the option value tests the value of the effort against its cost and the size of the prize tests the impact on the future of the company.

As for most strategic analyses the value estimate for the innovation strategy does not have to be correct to the final decimal point, but it should be good enough to indicate whether the initiative is really important to the company. It is a significant figure that puts innovation in the context of other strategic initiatives. With this estimate done by the strategy department, the feedback loop between innovation strategy and execution by the innovation team is closed.

The value of a portfolio of options

The main purpose of a portfolio analysis is to optimise the utility of the total innovation effort against the available resources in terms of budget and competences. The most common utility optimisation is to maximise the value of the portfolio, but this is not necessarily the best or the only one. Alternative optimisation objectives can be the best

The Value of Innovation

match with the in-house capabilities or the creation of a regular flow of novel product launches over time.

Assessing the total value of a portfolio of innovation projects is not straightforward. The value of the portfolio is not the same as the simple sum of the value of the projects due to the interaction between the projects. Not all projects can be as successful as assumed in the stand-alone assessment. Projects can be mutually exclusive or will compete for the same market space, certain projects will depend on the success of other projects and other projects will support each other in the market. Models to quantify this interaction between the options exactly are emerging, but they are still very complex. However, a manual correction on the total value of the portfolio is usually adequate to assess the potential impact of a portfolio.

Table 5.1 provides a simple example in terms of market share based on an innovation portfolio with three different power packs for the distributed energy market, for instance micro turbines, Stirling engines and fuel cells, and a small, efficient energy storage device. It is assumed that the three power packs each have their own specific markets, and, for instance each one on its own could have a market share of 15%. But, if they are all introduced simultaneously, they will also compete with each other and the total market share will be less than 45%. But all three power packs will increase market share if they can be offered in combination with a low-cost energy storage device, because this increases their efficiency and reduces overall cost. For the same reasons the market share of the storage device will increase in combination with a power pack.

The value of the portfolio will be less volatile than that of the individual projects and can give a fair impression of the potential contribution that innovation can bring to the value of the company.

Table 5.1 Effect of competition and synergy between projects
(numbers represent forecasted market share)

Innovation project	Stand-alone	With competition	With synergy
Power pack A	15	} Jointly 25	} Jointly 35
Power pack B	15		
Power pack C	15		
Storage pack	10		20

Insight in Innovation

But the innovation strategy should also be assessed in terms of risk. An innovation with a big impact and low chances of success needs careful management as such and in relation to other significant-risk-bearing operations. Innovation needs to be on the radar screen of the company's risk management team and quantified with the same metrics.

A useful result of a portfolio assessment in terms of (option) value and risk is establishing the 'efficient frontier' of the portfolio, as illustrated in Figure 5.3. The efficient frontier marks the best possible trade-off between reward and risk. The frontier can have only limited accuracy when there is much synergy or competition between the projects, but the frontier zone still provides a good reference point for assessing the relative attractiveness of the projects in the portfolio. Projects far removed from the frontier zone need to be challenged as to whether their call on resources is justified.

In practice, a portfolio needs to be assessed with a number of plots, such as contribution to the top line and the bottom line against risk, resource requirement, exposure and time to market. The choice of combination depends on the challenges and objectives of the innovation effort.

Most portfolio segmentations will not represent a true picture of the state of affairs, but suffer from one or more biases that tend to creep

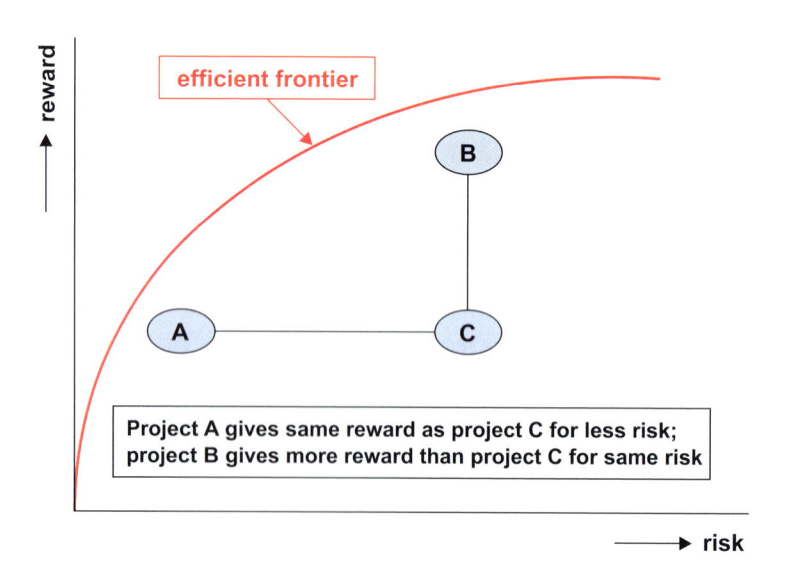

Figure 5.3 Trade-off between risk and reward in a portfolio.

The Value of Innovation

into the evaluations. A plot of the projects against time will often be skewed to the short and long term with a thin section in the medium term. One of the reasons is that, until the commitments to invest have been made, the length of the implementation stage is hard to estimate because it is for the most part out of the control of the project leader. Thus, before they have passed the valley of death, projects tend to be put into or close to the long-term and once they have passed the valley of death they are put into the short term.

Also the high-value projects tend to be in the long-term domain and the low-value projects in the short-term one. The reason is that the project leader will look at the potential scope for the value of the project as long as it is long term, but close to launch the estimator is also entrepreneur-to-be and knows that his estimates can become targets and he will tend to estimate the value of the project on the low side to increase the chance that he can meet his potential targets. Figure 5.4 illustrates the biased portfolio based on 'fair and honest' assessments by the project leaders.

Another typical weakness of assessment methods on future income is that they tend to be optimistic on the value side and conservative on the cost side. Therefore, a crosscheck on both the value and the cost side is important. The total value of the innovation portfolio should be in line with the value of the strategy set by management for innovation, and if this is not the case a serious rethink is required.

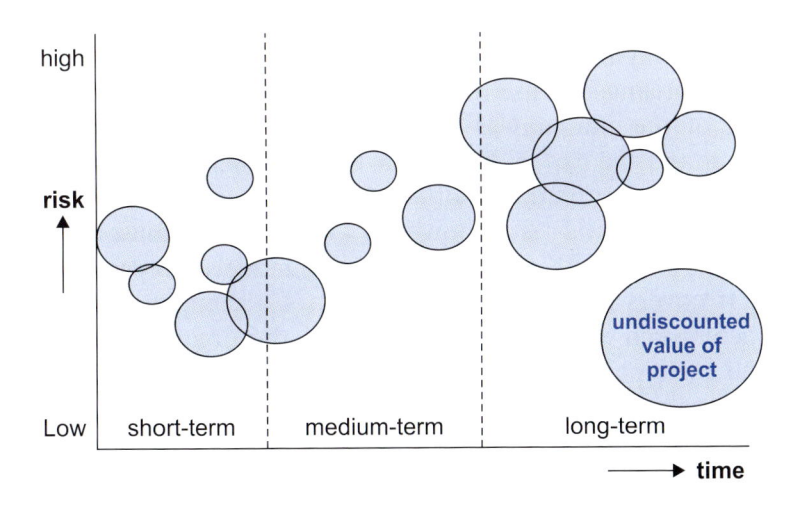

Figure 5.4 A 'naturally biased' innovation portfolio.

A fairly simple reason for a misalignment can be that the resources applied are too small compared with the strategic ambition and only incremental resources are applied to an innovation effort that has revolutionary changes as the objective. Sometimes radical innovation is treated, implicitly or unintentionally, in the same way as buying lottery tickets — as a chance of achieving great value with little commitment of valuable resources — and like most buyers of lottery tickets, the consequent small chances of success are conveniently ignored. The facts of life are that to create substantial value, substantial commitments have to be made. In 'hype-periods' it is possible to make lots of money with little effort, but the usual base line is that it takes time and effort to create major value. A single project can be exceptional, but a substantial portfolio will tend to bring cost and value within range of each other.

Innovation is not necessarily a low-cost option for creating new business. It ranks next to alternative options such as acquisitions and venture units, and all these options have high risks and significant cost. Too easily is the image of value creation by innovation set by the exceptional big winner, rather than in the context of the many mediocre performers and outright losers.

INNOVATION CREATES INTELLECTUAL CAPITAL

One of the intangible assets of a company is its innovation capability. Not many companies quantify the value of their intangible assets, which is surprising because for many companies the intangible part of the value of the company is more important then the tangible part. Innovation and the capability to innovate represent an important part of the intellectual capital (IC) of a company.

Intellectual capital is 'know-how with a recognised potential for value' as embodied in staff, documents, models, processes and customers. It covers all the intangible assets of a company and includes components such as the value of the staff, the organisational capabilities, the customer base, and the innovation capacity. Putting a value on the contribution of innovation is more relevant and more accurate if all the components of the IC of the company are valued on the same basis and in the same currency. Assessing the value of innovation in the context of the value of other intangible assets provides a better perspective on the company's intrinsic value and performance and

The Value of Innovation

gives innovation its proper position in the company's value creation processes.

Another interesting argument in support of the need for creating and using monetary figures for IC is that this type of information is a requirement of the post-industrial era [44]. Financial accounts came with the industrial era, for instance, the balance sheet was introduced in 1868, but the post-industrial era needs more and other data to manage and assess knowledge-based businesses. Financial accounts give data on cost of products, materials and labour, not on the value of intellectual assets and services.

A segmentation model for intellectual capital

The segmentation of intellectual capital is not standardised, but although the names differ from author to author [45–47] there is a fair degree of consensus on the main building blocks. Figure 5.5 gives a fairly standard example of the segmentation of intellectual capital, but each company needs to build its own model and segmentation to express its specific characteristics.

The word 'capital' has been used in the names of the components as suggested by Edvinsson, because we like to express the value of the

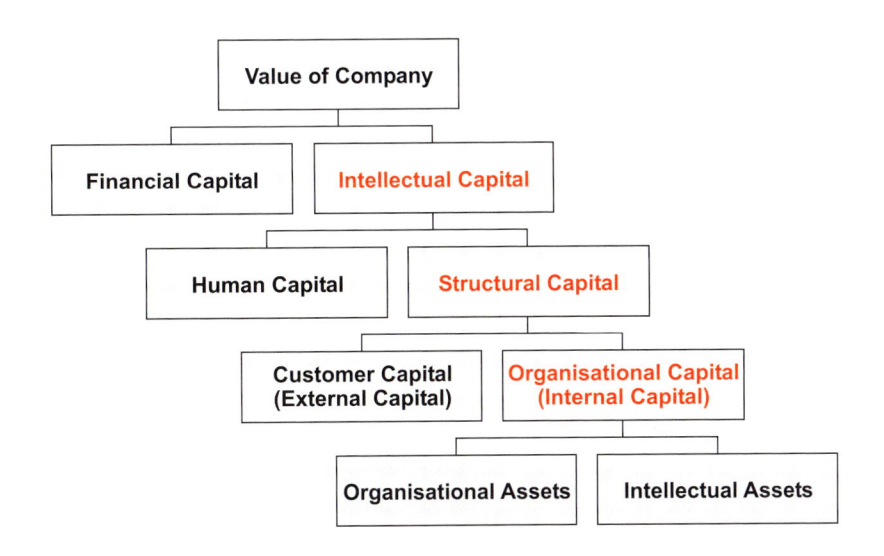

Figure 5.5 A model for segmenting intellectual capital.

IC in money terms, whereas Sveiby prefers the word 'structure' and Kaplan & Norton use the word 'perspective'.

The basic build-up of a model that can be quantified is to split a component into two parts, of which one can be determined and the other is split further, until a measurable or well-defined entity is created.

The total value of intellectual capital is derived from the market value of the company minus its net financial assets. Structural capital represents the residual value of the intangible assets of a company after office hours when the staff have left. Structural capital is split into external and internal components, either embedded in the customer relationships or in the organisational structures and processes. Organisational capital can then be split into its company-specific, constituting components; the intellectual assets such as the legally protected and the confidential know-how, the organisational assets, such as innovation and manufacturing capabilities, or certain specific and intrinsic competitive capabilities. Table 5.2 lists some of the more important components of intellectual capital.

The segmentation model illustrates why the accuracy of an estimate of the value of one component improves significantly if all components are estimated. The sum of all the IC component values provides an important constraint; the total value of all the components of intellectual capital has to be less than the company value. Whereas the estimate of each component by itself can be quite uncertain, the

Table 5.2 Constituting components of intellectual capital

IC segment	Constituting components
Human Capital	Tacit know-how
	Capabilities and competences
Customer Capital	Reputation
	Customer loyalty
	Contracts
Organisational Assets	Innovation capabilities
	Manufacturing capabilities
	Business models
Intellectual Assets	Retrievable confidential know-how
	Intellectual property (patents, trade marks)

The Value of Innovation

comparison with the other components and the limit to the total value greatly reduce the span of uncertainty.

Still there is a great deal of reluctance in business to adopt a routine quantification of the value of intellectual capital for all components. Some of the resistance is rather emotive; for example, the score on one of the prime performance parameters, the return on capital employed, reduces significantly. Other concerns need careful consideration; for example, valuable competitive information could be revealed if it were obligatory to publish detailed IC figures. The argument that the figures are rather uncertain is not a good reason to avoid establishing the value of intellectual capital. IC figures are not generated for use by accountants, but for use by management to better understand their business. The IC value figures need to be internally consistent, and relative values are more important than absolute values for managing the company with more focus. Absolute values would be required if inter-company comparisons were an objective. In that case agreement would be required on the definitions and build-up of all the components, but there is still a long way to go in that field.

The best reason for making regular, comprehensive assessments of the value of the IC is that several IC components such as patents and brand value are already quantified routinely by many companies, and all IC components are always valued, implicitly or explicitly, at transfer of ownership.

Several users of IC argue that intrinsically money is not the correct currency for measuring IC. Intangible assets need intangible metrics, such as the balanced score card, that are tailored to the specific value component so that they can lead to the correct insight and the proper action. In itself this argument is correct, but this holds for most value items in life and business. The advantage of money as a measure is that it makes communication so much easier that the reduced 'accuracy' is more than compensated. Money is the simplest and most effective language in business and is understood by all.

But caution in the creation and use of IC money figures is warranted. The reluctance of the CFO (Chief Finance Officer) to generate figures on the value of IC is understandable. The financial and intellectual capital statements represent two sets of figures expressed in the same currency, but with an intrinsic difference in quality. The accuracy of the financial figures is not necessarily much better than the IC figures, but they are produced via strict accountancy rules, while there are no rules or consensus on how to calculate IC figures.

However, if produced by the CIO (Chief Information Officer) the IC money figures can provide a useful management tool for identifying the value components in the company, for instance to validate the well-known statement used by many a CEO that: 'people are our most valuable assets'; many companies will not be able to substantiate that statement [48]. Creating an IC model helps the understanding of the value creation process by the company, and understanding the value of the IC components gives management an additional tool for assessing and steering the business.

The value of intellectual capital

There is no generally agreed quantification method for all the components of IC, although many methods for specific components have been published [49,50]. Management of intellectual capital tends to be qualitative and directional. However, in principle, it is no more difficult to estimate the value of the other IC components than it is for innovation.

At a high level of abstraction all methods try to establish a FMV (Fair Market Value), either on the basis of costs, income or market value. All methods still have one or more arbitrary steps in which a part of the total cost, income or value is allocated to the value of the IC component, with the allocation factors created by experienced professionals on the basis of limited data and seasoned judgement. The best-known example of the allocation factor approach is the Technology Factor method developed and used by ADL and Dow Chemical for valuing intellectual property.

In principle, the value-based approach is the most appropriate method for estimating the value of IC, but for most IC components information on market value will not be available. The exception is the total value of the IC; the total value of the IC of a company is easy to give and is defined as the total value of the company minus its net financial assets or market value minus book value. Similarly, for most of the IC components income figures will not be available because they do not generate income directly; the main exceptions are a few specific components such as patents that create a licensing income. Therefore, it is advantageous to take cost as the basis of the valuation methodology, because cost figures can be created for all segments. Starting with cost may seem quite inappropriate, because cost, price and value are different qualities determined by

The Value of Innovation

different mechanisms. However in most markets value and cost move in rather narrow bands relative to each other and in many industries and businesses the value/cost ratios are close to one, sometimes too close for comfort. The cost-based approach requires each IC component to be defined properly and comprehensively in terms of resources and activities. For instance, the human capital needs to include all cost to staff and the customer capital all the acquisition cost. Not all IC components will have good cost figures and a degree of judgement will be required to allocate the total cost of the company to the IC components.

Typically, cost figures for intangible assets are annual cost figures and they have to be transformed into capital cost figures. This can be done in a number of ways, for instance by estimating what the replacement cost of the asset would be.

The next step is to judge the value/cost ratios for the IC components; here it may be beneficial to sub-segment the components further to be able to use differentiated ratios. Table 5.3 gives an example of a set of ratios that can prevail in a specific business domain, but it is useful to create a specific one for the business at hand. The values obtained can then be crosschecked with alternative estimates based on income or market information where available and by ranking of the components.

Table 5.3   Example of differentiated value/cost ratios

Typical range of value-cost ratios	Characteristics of the conditions in the market place that support the potential for value creation
< 1.0	Market is disappearing; cost recovery becomes a problem
0.8–1.5	Product generally available (commodity); market is transparent. Competition is based on cost, and reward is on cost-plus basis
1.2–2.5	Market is protected or restricted; barriers to market entry exist. Product is differentiated and commands additional margin
2–4	Product is unique and in high demand. Alternatives are not readily available or easy to develop (the value of time advantage)
3–10+	Product is unique and protected; alternatives are locked out or do not exist

Insight in Innovation

This method has been applied to Shell Global Solutions for the conditions prevailing in 2000 and Figure 5.6 shows the result in a somewhat reduced and simplified format.

Figure 5.6 shows that comprehensive quantification is possible and delivers interesting results. Besides the familiar result that the financial capital is only a small part of the value of a knowledge-based company, it shows an interesting distribution of the IC. Typically, the value of the three main components – human, customer and organisational – is assumed to be of a similar size. For this company the organisational capital is dominant, with three major value components: the confidential know-how, the innovation capability, and the business model. This distribution is likely to be atypical because Shell Global Solutions was then a young company with a growing customer base and over time it can be expected that the distribution will become more balanced.

The value of intellectual property

Intellectual property (IP) is an important element in the innovation value chain. IP represents one of the knowledge components of a company and is part of its intellectual assets, defined as the explicit, or documented and retrievable, knowledge. The intellectual assets segment

```
                        'Market' Value
                            100%
                              |
              ┌───────────────┴───────────────┐
     Financial Capital                Intellectual Capital
          15%                                85%
                                              |
                        ┌─────────────────────┼─────────────────────┐
                 Human Capital         Customer Capital      Organisational Capital
                      25%                    15%                     45%
                                                                      |
                                                   ┌──────────────────┴──────────────────┐
                                           Organisational Assets              Intellectual Assets
                                                  20%                                25%
```

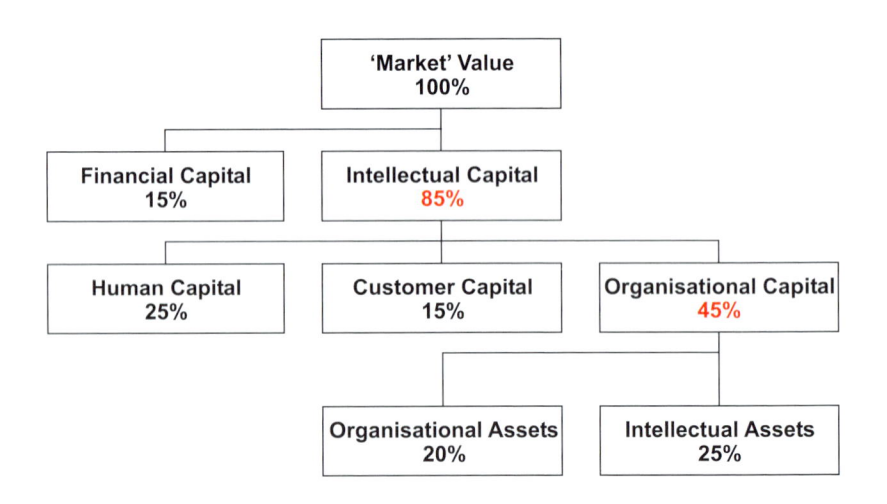

Figure 5.6 Build-up of the value of a knowledge company.

The Value of Innovation

has two main components: confidential information and legally protected know-how or intellectual property. The other knowledge component is tacit knowledge that is embedded in the staff and resides in the human capital segment.

The main reason for generating IP is to create a defence mechanism to safeguard that, later in the innovation supply chain, the value of the innovation can be monetised by the entrepreneur. IP is valuable because it can create freedom of action and prevent the competition from competing on the same terms. IP can create a competitive advantage by providing exclusivity and exclusivity can be used to defend high margins. Creating exclusivity is the main purpose of generating IP, but IP can still have value if, in the end, the innovation process is not successful and the new product is not brought to the market. In that case it may still be used as a strategic asset and sold to another company, as such, or as part of a technology swap; for instance, as one of the value items that a company brings into a strategic alliance. In addition to this use in strategic games, patents can be used to generate revenues in licensing activities. In general, these activities are lower added value on a per unit basis, but the total value can be large if the markets that can be entered in this way are big. Creating IP for defensive reasons only, for instance to limit the freedom of action of the competition, can have its merits, but the rewards are seldom high. Figure 5.7 visualises the basic options for monetising IP.

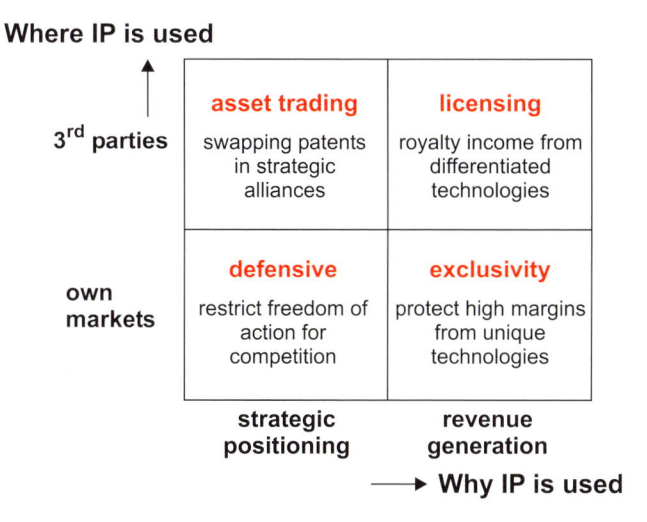

Where IP is used

	strategic positioning	**revenue generation**
3rd parties	**asset trading** — swapping patents in strategic alliances	**licensing** — royalty income from differentiated technologies
own markets	**defensive** — restrict freedom of action for competition	**exclusivity** — protect high margins from unique technologies

→ Why IP is used

Figure 5.7 The option matrix for monetising IP.

Usually IP is generated to create exclusivity, but for this purpose keeping the information confidential and exclusive to the company can be a good or even better alternative. This is so when the IP protects only one of several similar options for bringing a product to the market. If the issue is to limit freedom of action of the competition, then publication can be a low-cost alternative to a patent.

If, for one reason or another, the company can reach only a small fraction of the total market, it can be beneficial to license the technology to third parties and create value in that way. The licensing fees or royalties effectively transfer a part of the margins at the customer interface that are opened up to the licensee by the granting of the license to the licensor. For monetising know-how via third parties, IP usually is a necessary and efficient tool.

IP offers legal protection, but it has the intrinsic disadvantage that confidential information needs to be made public and this information can be used by the competition as a platform for creating an alternative approach. They can do this, for instance, by developing an alternative route to the market or by simply licensing the technology and competing in the same market. Disclosure of confidential information represents a reduction of intellectual capital and it requires active management of an IP portfolio to avoid patents destroying rather than creating value. IP management has many legal implications, but it is in the first place a business issue and an important part of innovation management that should not be delegated to patent attorneys or an IP department. IP are agents for doing business.

The IP value matrix given in Figure 5.8 provides an illustration of the typical values that can be expected for the different options for monetising IP relative to the cost of creating that IP. The comparison of the figures is based on the assumption that the third-party market is equal in size to the home market. If the third-party markets are much larger than the markets than can be accessed by the company that owns the IP, the top box options acquire much more value. Such a matrix can also be used to assess various IP strategies and to provide a basis for quantifying the value of the patent portfolio.

Each box in the matrix needs its own specific quantification approach. Patents in the licensing box (top, right-hand corner) can be valued rather straightforwardly on the basis of the licensing income and for the exclusivity box (bottom, right-hand corner) this can be done from the higher margins created through an exclusivity position. Patents in the other two boxes are usually not generated for that purpose, and effectively their value is zero until the specific moment

The Value of Innovation

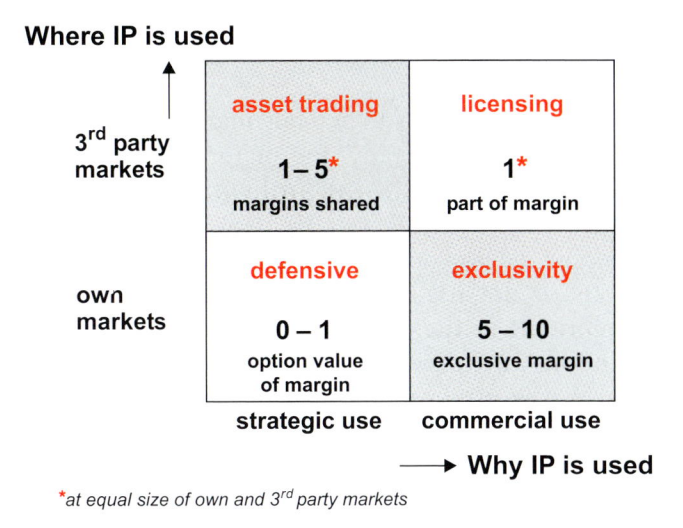

Where IP is used

	strategic use	commercial use
3rd party markets	asset trading 1–5* margins shared	licensing 1* part of margin
own markets	defensive 0–1 option value of margin	exclusivity 5–10 exclusive margin

→ Why IP is used

*at equal size of own and 3rd party markets

Figure 5.8   The value matrix for intellectual property.

that the strategic game is played out, and then the value of the patents in these boxes needs to be related to the specific opportunity. However, if the IP is generated for that specific purpose, an option value can be allocated to them based on the potential value of the transactions and the probability that the event will occur and when.

In general, the purely defensive use of patents is not enough for cost recovery. The reason is that the IP is only used passively and, if none of the competitors ever has the intention of creating IP in that domain, the IP remains valueless. But even if a competitor does intend to do it, the value of the IP would only be equal to the licensing fee that the competition could be charged. Defensive patenting needs to be scrutinised carefully; it is easily done, but it is a cost factor without an active value component.

An IP portfolio needs to be actively managed and quantification of the value of the portfolio is an important tool in that respect. The prime objective of IP management is to ensure that the patents in the portfolio are used in the right application. The match between the protection offered by the patent and the intended application is not only important at the drafting stage. It needs to be assessed regularly during the lifetime of the patent, because of changes in business strategy or in the competitive arena. Patents that are initially intended for use in margin protection may need to be used later for licensing purposes once competitive alternatives have entered the market and in that way extract some additional value from the IP. Or defensive patents are no longer needed, because there is no longer interest in pursuing that line of business.

The second main objective is to actively manage the cost of maintaining the patent portfolio. Typically, the major part of the value of the IP portfolio comes from only 5–10% of the patents and the cost of maintaining the other 90–95% can be high. Many patent portfolios suffer from benign neglect and patents are maintained without good business reasons or not used to the full extent. The probability that there is a Rembrandt in the company attic [51] is low, and not much better than finding an Edison in the boardroom [52], but it is good practice to check the attic regularly. It is even better practice not to store paintings in the attic.

SUMMARY

1. The value of innovation needs to be determined from three perspectives: the potential value that the customer might be able or willing to give, the value of the strategic intent and the value of the projects in the innovation portfolio.

2. The value of an outside-the-box innovation project is best assessed as an option value to reflect the risks and uncertainties during the development process, the possibility to create alternative development options and the value of future management decisions.

3. The innovation capability of a company is part of its intellectual capital. The relevance and accuracy of estimating the value of innovation is enhanced if all the components of intellectual capital are valued on a similar basis and expressed in the same currency.

4. Creating a model for the segmentation of intellectual capital helps in understanding the value creation process in the business, and knowledge of the relative value of the IC components gives management a tool to understand and steer the business.

5. Intellectual property is a powerful tool in the innovation value creation process and IP can be very valuable. But IP needs to be actively managed for that purpose. Creating IP brings exclusive know-how of a company into the public domain and, if left to that, it creates only costs and a loss in value.

– 6 –
Sustainable Innovation

Sustainable development is a wide-ranging concept that is still in its early stages of development. It involves many stakeholders with different perspectives and over time several definitions of sustainable development have been given. The first definition dates from 1987 and was given in the Brundtland report Our Common Future:

> *"Development that meets the needs of the present, without compromising the ability of future generations to meet their own needs."*

It represented a breakthrough in the thinking about the development of our planet by including future generations as stakeholders. Later definitions such as that given by the United Nations in 1992,

> *"Improving the quality of human life while living within the carrying capacity of the supporting ecosystems,"*

and the World Council for Sustainable Development (WBCSD):

> *"A desire for greater equity, quality of life and environmental well-being, today and for future generations"*

introduced other important elements of sustainable development such as quality of life, social equity, carrying capacity of ecosystems and environmental health. These elements are important aspects of sustainability that need to be recognised and included in the assessment of innovation for sustainability.

SUSTAINABILITY AND INNOVATION

The link between sustainability and innovation is direct and significant; both are oriented towards change and the future. Sustainability is about concern for the well being of our future, and innovation is about creating new products that can generate value only if they fit in the future. A good antenna for what the future may or may not need or bring is an important asset for successful innovation, but it is also essential for sustainable development. The world as is appears to be in a non-sustainable mode of operation and we have to change the way we produce and consume our products. The vehicle for changing the world towards a more sustainable future is innovation, as the tenth law of innovation expresses:

Law X   The purpose of innovation is to create desired, valuable change

Law X may sound a bit obvious and not too profound, but it is important to realise that there is no innovation without change and that innovation is the key mechanism for creating change that is intended and needed. Change that occurs without positive intention usually leads to erosion or destruction of value.

Much commercial innovation is initiated not to create change, but with the implicit purpose of creating more of the same, be it profits, revenue or growth. But innovation does not bring more of the same but change, and outside-the-box innovation brings radical change. Law X expressed negatively as '**if you don't want change, don't innovate**' makes the point that there is no need for innovation if things are just right the way they are. If a company, a country or a customer is not interested or prepared to change, innovation cannot be successful. An innovative company has to be flexible.

In a non-sustainable environment change is a necessity and thus in today's world there is no choice but to innovate. Innovation has to support sustainable development and become sustainable innovation to create the required and desired improvements in the quality of life within the carrying capacity of the ecosystems of the planet. Sustainable innovation needs to be "*environmentally sound, socially acceptable and economically viable*".

The challenges for sustainable innovation

Innovation as such is not easy, but sustainable innovation has to meet three additional challenges; it has to meet societal expectations,

Sustainable Innovation

create an equitable distribution of value along the value chain and fit within the carrying capacity of its supporting ecosystems. Sustainable innovation requires benign technology and business models to meet the needs of all stakeholders and the resulting 'change' has to be in tune with society in order to be successful. These three requirements for sustainable innovation will each be discussed in more detail.

1. Sustainable innovation must create options that fit in the prevailing value system of society

'Sustainability' can be seen as the emerging core value of society, just as 'progress' was in the industrial era. In the industrial era technology was seen as the engine of progress, and society's basic attitude was positive towards technological developments. But after 200 years of industrial progress, the legacy is an ambivalent and sceptical appreciation of technology [53]. The attitude towards technology is a key challenge for sustainable innovation. In terms of environmental impact technology is seen both as the cause and as the potential solution; a paradox first expressed by Gray [54].

'Sustainability' as the new key concept representing today's societal driving value does not simply replace the notion of progress from the industrial era, but modifies and extends it. Progress in a sustainable world not only has to be good for today's society, but also for tomorrow's'.

Societal acceptability always played an important role in game-changing innovation. In the industrial era an innovation had to be perceived as contributing to progress; in today's society innovation will be judged on its contribution to sustainability. There was resistance in the industrial era to many types of technological progress such as textile machinery, the train and the car; not everybody saw electricity as a blessing and some people still think that the world would be a better place without television. Resistance to change has always been part of the diffusion of innovations into society. To avoid or minimise these types of conflicts, innovation has to start with societal needs and the dialogue with society needs to be part of the innovation process.

But just as progress was never a generally agreed concept, sustainability is and will remain a disputed vision. There may be some degree of consensus on the long-term vision at a sustainable world that is prosperous, safe and clean, but there will be a wide range of opinions on the priorities and different stakeholders will have different perspectives. Some will emphasise social equity and poverty, others a clean environment, deforestation or climate change, not to mention wars and diseases.

Furthermore, sustainability is a moving target; not fast moving, but it is not static. Today's concerns and values will be different tomorrow. Priorities will shift because certain problems are solved or become less urgent. Since the publication of Silent Spring [55] almost 50 years ago, environmental concerns have moved from pesticides to smog and radiation and further to acid rain, ozone depletion, global warming and genetic modification. Ecological priorities have their own lifecycle. There is a inherent shift in environmental issues with increasing wealth, starting from local problems such as clean drinking water and sanitation, to regional problems such as air and water pollution, and finally to global problems such as climate change.

Thus even with a global desire for sustainability, priorities in different parts of the world will differ and these differences need to be recognised as genuine by all stakeholders [56]. In the rich consumer markets choices for the customer and global environmental concerns will rank high; in the emerging economies growth without the negative side effects of the industrial era will be a prime objective and in the survival economies the supply of basic needs is top priority. Each set of problems will need specific and dedicated innovation; cascading the solutions from the past from the developed to the developing countries is not a sustainable solution.

2. Sustainable innovation must create equitable value for the customers and stakeholders along the value chain

For most of the population in the world economic growth and creation of wealth have the highest priority. On a global scale, social inequity is the key hurdle to sustainability and building equity into the value chain of new products should be a key challenge for sustainable innovation. There are no easy solutions for the equity problem. But it is important to recognise that for this issue technology cannot bring the solution; it may do so on the environmental frontiers, but not in the area of social inequity. Equity is in the first place a political issue, but business can make a contribution.

There is widespread opinion that a sustainable world will need a fundamentally different way of producing and consuming our products, including a different way of using and valuing natural resources [57–59]. This could be an attractive and a sound way of tackling sustainable development, but the problem is that this approach does

Sustainable Innovation

not work yet. Society at large is not committed to sustainable development to that degree and is not willing to pay that price. A minority of consumers is willing to pay a price for sustainability, but most feel that sustainability is a concern for producers and authorities and that sustainability, like quality, should be free and an integral part of any product. Similarly, many businesses are afraid that sustainable development creates an un-level playing field for them and are reluctant to enter the game.

On the issue of 'who will pay for sustainability' the priorities of stakeholders differ most and this aspect of sustainable development creates a major challenge for sustainable innovation.

3. Sustainable innovation must fit within the carrying capacity of its supporting ecosystems

An overarching objective of sustainable innovation is to reduce the footprint of human activities. The human footprint includes the use of natural resources, infrastructures and space; the generation of waste, losses and emissions; and visual, smell and noise pollution. How far the human impact has to be reduced is still open for discussion. A zero impact is impossible, but is 'no negative impact' or 'within standards' good enough? Many observers believe that improvements in the current technological systems will not be enough and a total replacement or creative destruction is required to make the footprint small enough for our planet [60]. Whether total replacement is required is open for debate, but it is fair to assume that only gradual improvements to the current systems will not be enough to turn the human footprint on the ecosystems from 'red and too big', to 'green and fitting'. A step change in the size of the human footprint is required as Figure 6.1 attempts to illustrate. Incremental, inside-the-box innovation will at best deliver an amber footprint. The world needs genuine green solutions and green solutions need outside-the-box innovations that deliver a step change in environmental impact. In the area of reducing the human footprint technology needs to make a very important contribution.

In addition to reducing the footprint of human activities, sustainable innovation should aim to increase nature's carrying capacity or at least restore it to its original capacity; Figure 6.1 attempts to illustrate that effect as well. Mismanagement of natural resources has reduced the carrying capacity of the global ecosystem; well-known examples are deforestation, over-extraction of water from rivers and lakes for irrigation, monocultures and over-fertilisation, water and air pollution. But just

Insight in Innovation

Figure 6.1 The human footprint needs a step change.

as global food production and the carrying capacity of cities have increased by orders of magnitude over the last centuries, it must be possible to increase the capacity of the natural ecosystems. This approach to a more sustainable world has received less attention than the reduction of the human impact, but could be a very fruitful domain for innovation.

However, whereas impact reduction may suffer from the technology paradox, this approach may suffer from the management paradox: *human management is both the cause and the potential solution*. Good management of our natural resources is required to restore nature's natural carrying capacity. And, although technology has to deliver a key contribution, Grübler's statement [61]:

> *"only technology can liberate the environment from the consequences of human interference,"*

is too optimistic; technology alone will not be enough. Sustainability is a pervasive problem and needs contribution from all aspects of economy and society to be resolved.

Innovation that aims to increase the carrying capacity of the ecosystems creates an opening for a more positive valuation of technology

Sustainable Innovation

and a more optimistic perspective on the future. Progress throughout history has been based on innovative technologies. From the discovery of fire through the agricultural and industrial revolutions, technology has been the agent for change and associated progress. The history of our planet can be seen as a series of step changes, with each revolutionary change based on a new set of disruptive technologies responding to new societal values, which in turn created a new economic and cultural climate. Figure 6.2 illustrates these revolutionary changes with the effects on global population and climate.

Each revolutionary transition also brought unintended side effects that sometimes took a long time to remedy. The social misery in the earlier phases of the industrial revolution and the subsequent degradation of nature's eco-systems are well-known examples. The digital revolution should bring both a richer world economy, and human activities that are in balance with the global ecosystem. This optimistic perspective may balance the somewhat negative outlook on the future of

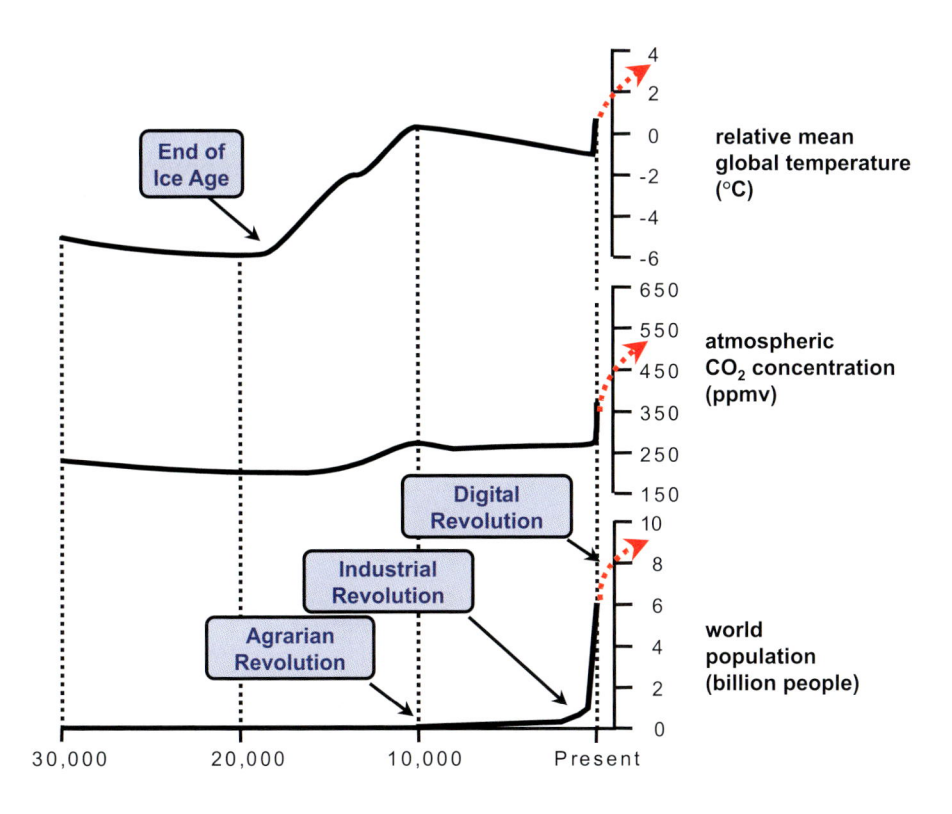

Figure 6.2 Revolutionary global changes.

Insight in Innovation

the world by many concerned environmentalists who see a significant reduction of consumption as the only remedy [62].

The sustainable innovation model

Balance is the key word in sustainable development and innovation. In the classic cascading model for innovation technology was the main driver, in the bridge-building model business and technology were the two drivers, but in a model for sustainable innovation society needs to be a driver as well. In addition to technology and business, societal values have become an independent driver for change. These three engines for change—business, society and technology—need to be in balance and interact effectively in order to create sustainable change. Sustainable innovation needs a new model that reflects the required interaction of its three drivers. The two-component, bi-modal bridge model has to be extended to a three-component, tri-modal system. Figure 6.3 shows a model that uses 'values' as the interacting agent between the three engines for change.

In the model the cycle for sustainable innovation begins when society starts to adopt or develop new values and reject certain products in the market place. Simultaneously, technology is stimulated to create new products that fit the new values, which in turn can be brought to the marketplace by business. The reject mechanism by society as a result

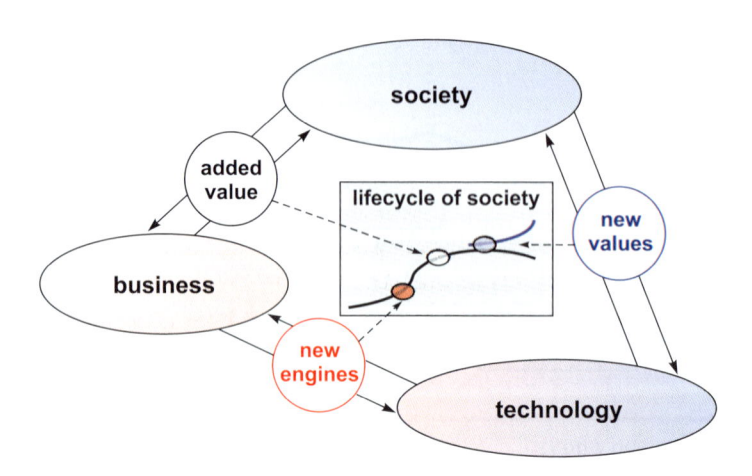

Figure 6.3 The value-driven model for sustainable innovation.

124

Sustainable Innovation

of the new values is a key element in the innovation model. This mechanism was not sufficiently present in the classic cascading model of the industrial era – the customer had little choice – and in the bridge-building model the mechanism was not operational, because society and business shared the same core value: the creation of wealth. Unconstrained extraction of resources at all levels to fuel the economic growth for today as prevailing in the industrial era, is being challenged by concern for future generations that also need a living planet.

At a cultural level the role of science and technology can be described as that of a creator of new options in response to societal aspirations, and sustainability can be understood as a societal value that creates a demand for new technological and business values and strategies. Sustainability is a societal 'driver', complementing marketing 'pull' and technology 'push'. It should be noted that all engines for change operate at all times, but at certain times a specific engine is dominant and sets the scene as indicated in the inside box in Figure 6.3.

One could say that in the pre-industrial era commerce was the main driver for change, that technology created the industrial era, and that societal drive has to create the sustainable world. The eleventh law of innovation expresses the requirements for sustainable innovation:

Law XI   Sustainable innovation has three value drivers — technology, business and society — that need to be in balance to create new choices for the customers of today, without compromising the options for the future

In sustainable innovation the three engines for change operate in harmony and the value drivers are in sync. For a company this means that there needs to be an openness to society to pick up the signals of change and a willingness to participate in the public debate on the desirability or otherwise of intended changes. Sustainable innovation is interactive innovation. Whether a company wants to convert its innovation effort into a sustainable innovation effort is, up to a point, a matter of choice. The interaction with society at large can either be seen as an unnecessary time delay or as a mechanism to increase the probability of success. In the long run, however, sustainable innovation is needed for a company to survive in the evolution of society. Sustainable innovation will make a company more resilient to the emerging large changes in its environment by providing options to choose from that

SHELL AND SOCIETY

In the first half of the last century distribution of wealth was a key concern and Shell focused on delivering quality products to all customers in a world where quality was not a given. It very successfully used the famous slogan 'You can be sure of Shell' to communicate this message to its customers.

After World War II the core societal concern was economic growth and the creation of wealth, and the Shell focus became the supply of sufficient energy to the world, with 'Shell Helps' as the new message to the customers. With increasing wealth the Shell proposition to the customers evolved in line with societal expectations and over the years additional features were included, such as convenience at the service level and cleanliness in the products. In today's world sustainability is added, but the new slogan to communicate this convincingly has not been found yet.

Shell used to be leading in responding to changes in societal values. However, in the last decade of the twentieth century, Shell found itself late and lacking! Shell was good and confident in the 'Trust me' world, but not ready for the 'Show me' world. The value shift in society was not recognised in a timely fashion and built into the business values, see Figure 6.4. Fortunately the events with Brent Spar and Nigeria were used as valuable learning points and triggered a strong and sustained response. The decision-making process was changed from the DAD to the DDD approach; from Decide, Announce, Defend to Dialogue, Decide, Deliver [63].

Sustainability is now an integral part of Shell's business principles, dialogue with society is part of the business process, and the earlier perceived conflicts between economics and sustainability are being resolved. Innovation is an essential part of the process to move away from *being part of the problem*' to *'being part of the solution*'.

The new strategic direction in Shell for innovation aims at paradigm changes that can offer sustainable solutions to society. With global warming as a key concern, this paradigm shift towards sustainability is even more pressing for the hydrocarbon industry. From this perspective the investments made by Shell in Renewables and Hydrogen can be seen as forms of sustainable innovation, with being prepared for a different future as the key driver.

make these changes either less painful or more attractive. It is about survival of the 'fittest' and in tomorrow's world the fittest may well mean the most flexible and innovative. Without sustainable innovation the company has no options with which to respond to the changes and may be forced to just follow the course of events, whether the end result is desirable or not.

Sustainable Innovation

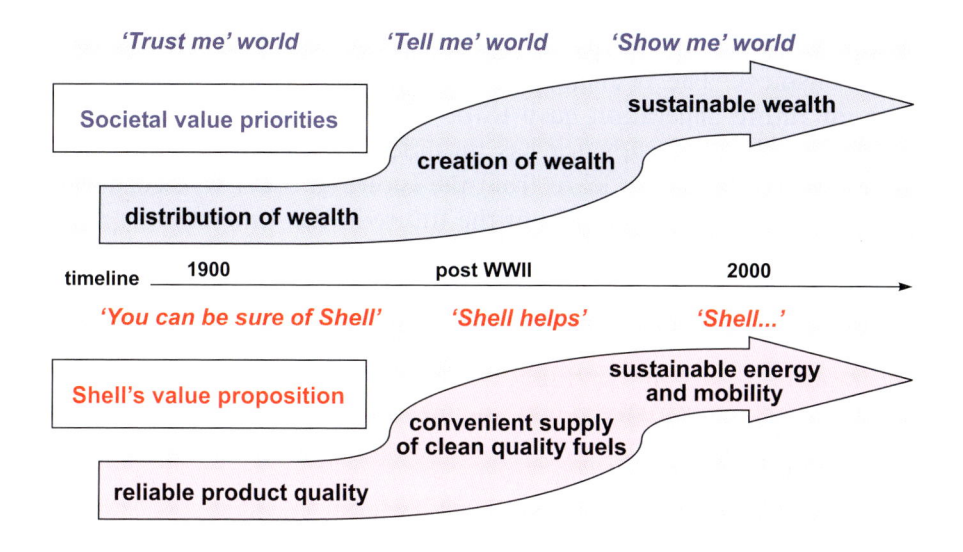

Figure 6.4 Shell's responses to societal shifts.

It might be illustrative to put the changes required in a company in order to stay in business into a historical context. The experience of Shell as a global company may serve as example. The Shell history of over one hundred years is just enough to see the changes from the industrial to the post-industrial era and to explore the impact of sustainability on innovation.

When one looks back, one can see Shell responding over time to changing societal needs in the way it delivered its products to its customers. Sustainable innovation needs to be business-driven, technology-enabled and socially acceptable. When the issue is adaptation to a changing environment, technical innovation by itself is rarely sufficient, but has to be accompanied by commercial and societal innovation. It is not sufficient to change the offering to the customer; the interface and interaction with customers and society have to change as well. Sustainable innovation aims at creating new choices for the customer in tune with society.

LEARNING FROM THE FUTURE

Sustainable development should be the innovator's dream. It creates a need for change based on new values and the innovator has to express these new values in new technological applications and new ways of doing business. The challenge in sustainable innovation is that the needs of future generation have to be included in the process. To be able to do that, the innovator has to understand what the future may bring or need; he has to learn from the future in order to incorporate the interest of the stakeholders of the future in the innovations.

Trends

An important tool in learning from the future is the analysis of trends: trends in customer values, in emerging technologies, in business models. An essential part of trend analysis is to look for potential discontinuities in the trends to be aware of possible threats and break-out opportunities. A very important discontinuity is the emergence of a breakthrough technology that allows the creation of new clusters of technologies and a whole new range of applications. A new cluster of technologies gives rise to an abundance of innovations and sustainable innovation can create an advantage for itself, because it will express the future values and expectations of the customers and have the emotional edge to start with.

Trends run from history into the future. Understanding the future cannot be done without understanding the past. Analysing historical developments is an integral part of creating future images. History provides a rich mixture of examples of how complex systems such as businesses and societies have responded to innovation and change. The technology of the future has to express 'sustainability' as the societal value of choice. What this means is hard to predict, but looking back at the industrial era we can try to understand the links between the value 'progress' and the industrial technologies and attempt to translate this back into the future for the value 'sustainability'.

The leading technology of the industrial era was mechanical power, which replaced animal power as the extension of human power. In the coming era digital power may well be the leading technology as the extension of the human intelligence. Associated, parallel changes may come from a shift from 'making things' to 'doing things', from manufacturing to servicing, from owning to experiencing. All these trends fit in an overall trend towards dematerialisation.

Technologies that may do well in such an environment are, in addition to the pervasive and ubiquitous information and communication technologies, the micro-technologies that have the potential to reduce

Sustainable Innovation

the footprint of the manufacturing industry, the experience industries and the energy industries. Fossil energy can be seen not only as the power source of the industrial era, but also as its symbol and therefore, by reverse logic, destined to become an energy in decline in the post-industrial era. A new era needs a new symbol and a new power source. The high-temperature, high-pressure, high-power technologies that achieve their optimum performance at large scale may need to be replaced or complemented by smart, bio-catalytic or electro-chemical, small-scale processes that extract advantage from the economy of numbers and from their use close to the customer for high efficiency and minimal waste.

With a similar logic, one could argue that the fuel cell car will not replace the ICE-car unless the VROOOM factor, which is a power-related factor of the industrial age, is replaced by an emotional factor that fits the digital age; this factor could for instance be the 'smart (digital) power' of the all-electric car.

Table 6.1 indicates, by way of tentative example, a few potential shifts in values and technologies that are relevant for sustainable innovation. Such shifts do not necessarily mean that the old technologies disappear, but they lose their momentum and dominant position. This table builds on the famous work by Toffler [64] on the waves of technological change.

Other authors and institutes with different logic and methodologies have made similar forecasts on technologies that would fit in a

Table 6.1 Shifts between innovation eras

		Industrial era	Digital era
Values and characteristics		Progress	Sustainability
		Mechanical power	Embedded intelligence
		Making and owning	Doing and experiencing
		Division of labour	Connecting knowledge
		Economy of scale	Economy of numbers
Key technologies		Construction materials	Intelligent materials
		Power from combustion	Electro-chemical micro-power
		Chemical processes	Bio-conversion
		Pharmaceuticals	Life sciences
		Physical mobility	Virtual mobility

sustainable world. The OECD [65] mentions advanced sensors, biotechnology, clean car technology, product recycling, water treatment, waste treatment, micro-manufacturing, renewable energy and photo-voltaïcs. According to Dearing [66] the fundamental challenges for the success of the emerging technologies are given by the existing infrastructures, their economics and the slow acceptance by the public of new approaches. Other authors see the world today in a period of structural discontinuity where old mature value and infrastructure systems are replaced by new ones in 'gales of creative destruction' and the take-off of a new Kondratieff cycle [67] as the start of a long period of economic growth [68].

All these cycles and transitions have very long time periods. Changes in techno-economic values may take hundreds of years, the lifecycle of an infrastructure typically is around a century, and the Kondratieff cycle is about 50 years. Most major innovations take a long time to diffuse into society and the slow initial part of the S-shaped development curve for game-changing innovation can take 20 years or more before the fast take-off starts. However, the mobile phone and the Internet are examples of extremely rapid diffusion and may be in the digital era innovation can go faster. In general, the first application of a new technology is the slowest to diffuse and any innovation strategy for the leader has to include a plan of how to stay ahead of the 'fast follower' in the market place.

Techniques

Besides learning from trends and discontinuities there are several other ways to learn from the future and use these insights for innovation.

Shell is well known for the use of scenarios as tools for understanding the future. Scenarios are alternative stories of how the world may develop. They are not predictions or forecasts, but credible and challenging alternative stories that help to focus on critical uncertainties and to explore 'what if'. For innovation, scenarios are very useful for understanding which important elements in the future could transform the business or the conditions that are required to implement potential innovations.

Backcasting is an alternative process to learn from the future. In this method, a desirable image of the future environment and the position of the company in that environment is created and then the possible pathways to achieve those images are explored. Backcasting may seem

Sustainable Innovation

to be a very appropriate mechanism for sustainable innovation, because the process could start with an image of a sustainable future. The danger is that the image fixates the future, whereas societal values shift and the future remains unpredictable. For backcasting to be effective there must be a reasonable degree of control on the ways and means to achieve the desired future; this may be the case for governments in certain areas, but is typically not the case for companies.

Technology road mapping tries to predict the next generations of technology based on extrapolations in science, technology and applications. Technology mapping has been very popular. The danger of technology mapping is that the developments seem to have an intrinsic logic and thus a high credibility, while in reality the development of technology has a much more random path. It is very risky to predict more than one technology generation ahead.

ASSESSING SUSTAINABLE INNOVATION

Both technology and business have key roles to play in sustainable innovation, with new business models addressing the equity issue and technology the impact on the ecology. Assessing whether a business model is equitable can be done ultimately only by the customers along the value chain, provided that they have choices and there is competition.

Assessing whether a technology is sustainable is not straightforward. There is no technology with an application that has 'zero negative impact' on the environment. Most of the time a new technology offers advantages in one respect and disadvantages in others. The impact on the environment has many dimensions that are not always easy to measure or compare. For a number of negative impacts on the environment authorities have set standards, but meeting these standards is not automatically equivalent to being green or sustainable. And standards can change in time as a result of new insights or priorities. What is green today can be grey tomorrow. An innovation may be acceptable as a single application, but can create major resistance when applied at large scale. Ultimately, a new technology has to meet the emotional value system of society, and, being hard to predict, this adds another risk factor to technology-enabled innovation.

The consensus vision on the requirements of a sustainable technology includes elements such as, ultra-low emissions, recycle of products, and cascading of waste and by-products as feedstock for downstream technologies. Sometimes the image of the tropical jungle is used to describe the sustainability aspect of such a world built on sustainable technologies. In the jungle nature realises a system where 'nothing goes to waste', but everything is cascaded within a closed cycle. This is a powerful vision, but it has an intrinsic element of conflict for the industrial world. A system that is fully cascading and recycling will create a very rigid infrastructure for itself, because the links between the elements in the system become exclusive and inflexible. Such a global fixed infrastructure would become the ultimate hurdle for innovation and change, and such a static world would be totally different from the way of life in the dynamic world that we live in now.

This possible conflict is for the moment primarily of theoretical interest, but this excursion into the value system of the future is important for increasing awareness that value systems and priorities will shift over time. A fixed, static vision on sustainability is not sustainable. The image of the jungle as a model for a sustainable world is not adequate and needs – as a minimum – to be enhanced to a continuously evolving jungle.

Criteria for sustainable innovation

It is customary to assess sustainable innovation on a triple basis: economic, ecological and societal. Effectively the last two criteria are needed to include the externalities in the assessment, whereas, traditionally, innovation is assessed primarily against economic and company-centric criteria. To assess sustainable innovation, the set of screening criteria for innovation projects as given in Table 3.2 has to be expanded to include the criteria for the externalities as done in Table 6.2 for a few criteria by way of example. The framework and the screening process for innovation can remain the same. The screening criteria in Table 6.2 for sustainable innovation have been arranged into two groups: company internal and external. The six sets of criteria are generic, but the individual criteria are specific to the company and the innovation domain, and have to be specifically defined for each stage as discussed in chapter 3.

The screening criteria for innovation should assess the value of the innovation project in three ways: to the company, to the environment and to society. The screening criteria should be pragmatic and no more

Sustainable Innovation

Table 6.2 Screening criteria for sustainable innovation

Internal criteria	External criteria
Strategic criteria	**Equity criteria**
– strategic fit	– fair business model
– value for customer	–
–	**Ecological criteria**
Technological criteria	– ecological fit
– fit for purpose	– emission level
– intrinsic safety	– recycle and cascading
–	–
Economic criteria	**Societal criteria**
– value to company	– societal fit
– profitability	– eco-certification
–	–

sophisticated than our knowledge allows. This will mean that a number of the criteria will be more qualitative, because it will not always be possible to accurately quantify the impact on the eco-systems or the positive interaction with society. The suggested use of 'fit criteria' for ecology and society is based on the widespread and successful use of the 'fit'-criterion for strategy. In general, there will not be an ecologic or societal strategy available to judge against, but the 'fit criteria' provide a sound common sense basis for an overall assessment of whether the idea will be able to meet environmental and societal requirements.

Each innovation will have to fit in its own eco-system and when the fit is poor there will be resistance, losses and waste of energy. For instance, an IT innovation in Silicon Valley has a more natural fit than a mining activity in the Arctic. The 'fit criterion' for strategy can often be translated into the question whether the company would like to play the game if the innovation were to be successful. For ecology and society, similar translations can be made to test whether nature or people would be happy to play the game.

There is a danger that with too many criteria, too many ideas are killed prematurely. It is important to define and select the right criteria for the purpose with sound judgment. The primary objective of screening is to find the winners, not to make sure that there are no losers. In principle, the screening criteria in Table 6.2 only provide yes/no answers as to whether the projects meet the sustainable innovation criteria and thus can be economically viable, socially acceptable and environmentally sound. This screening has to be followed by a portfolio assessment to create the best collection of projects within the available resources. In the portfolio assessment, maximum value generation has to be balanced against a good distribution with respect to risks, timing, call on resources and capabilities.

Sustainable innovation needs to be anchored in the company to make it a sustained effort. Sustainable innovation needs:

- To be in tune with societal developments in order to anticipate future values and needs
- Sustained commitment to innovation by embedding it in the company strategy and adopting sustainability in the company principles
- A sound management process based on a deep understanding of the innovation process and on the needs and expectations of the stakeholders
- A strong external orientation to create diversity of perspectives and awareness of customer needs and values
- Leadership from the top and drive from the bottom to create a focused momentum.

Sustainable innovation creates societal capital

The shift in the value system of society resulting from the drive towards sustainable development will have an impact on the value of a company. Sustainable companies should create a premium value in the markets. It is too early to say by how much and in which way sustainability must be valued, but early indications are that companies that score high in indexes for sustainability also do well in the stock markets. Products and services that are appreciated as green by the customers could increase in value as well as in their life expectancy. On the other hand, old processes and products that have a big negative impact on the environment could become obsolete earlier than would otherwise be the case.

The change in the value of the company in the context of sustainable development equates to a change in the value of the intellectual capital. Participation in sustainable development by a company creates a specific form of intellectual capital, what we will call societal capital and which is part of the intellectual capital embedded in external stakeholders, just like customer capital, see Figure 6.5.

Sustainable Innovation

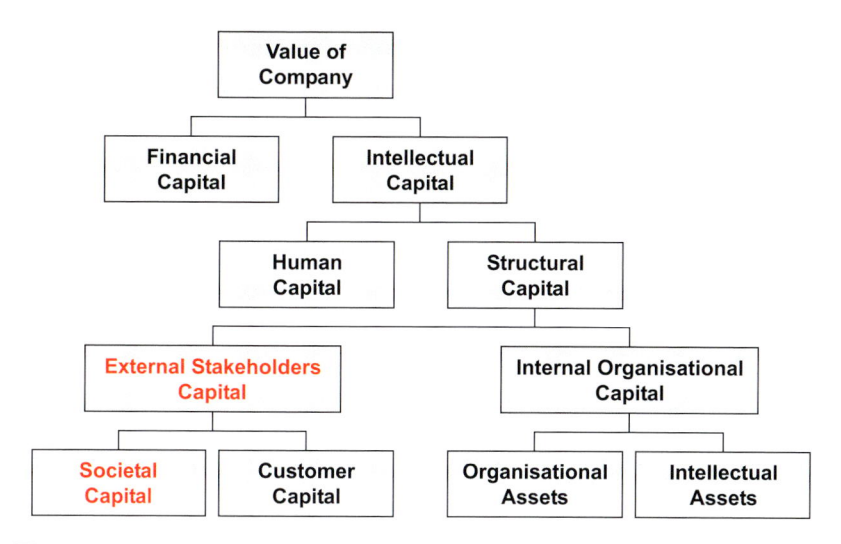

Figure 6.5 Sustainable innovation creates societal capital.

Societal capital represents the credit given by society to the company. The main components of societal capital are the licence to operate and the licence to grow. The company has to earn the right to these licences from the value of its products and services to society and in the way it lives up to societal expectations. Part of the value of the reputation of the company belongs in the societal capital box. It could be defined as the passive part of the reputation, not based on active customers but on other stakeholders that in one way or another can influence the decisions of the customer, as can be done by political boycotts or other negative campaigns in the public domain. Outside-the-box innovation is an essential contributor to societal capital, because it is the key factor in responding to changes in society, in providing resilience and supporting continuity to the company. In the words of Darwin:

> *"It's not the strongest who survive, but those most responsive to change".*

Several components of IC will become more or less valuable when assessed in the context of sustainable development. It will take time before these effects can be quantified with any accuracy, because the impact of sustainable development in the market place is not clear yet. Most of the impact will be in the structural capital. Besides societal capital, i.e. reputation, innovation capability may be one of the components of IC that increases most in value.

Sustainability may also have an important impact on the use and value of intellectual property. Trademarks may generate a big premium when they embed the sustainability value. In a functional sense, patents may become less relevant in the digital era because they protect new ways of making things rather than new ways of doing things. In the digital world value is created by increasing the number of users rather than by limiting the number of suppliers. Technological means to protect margins by embedding the proprietary technologies in the applications can become more effective than legal means in the form of patents. In a way this technology-based approach is similar to what was done in the pre-industrial era when know-how was embedded in the 'master' and only disclosed via the master–pupil system. But this reduction in functional value of patents to support innovation may well be overwhelmed by the increasing use of patents as an instrument to create value via legal ways and means.

SUMMARY

1. Innovation is the vehicle for change towards a more sustainable world

2. Sustainable innovation aims to create new options to reduce social inequity in the global economy as well as the impact of the human footprint on the ecology. Priorities in sustainable development depend on the prevailing societal needs and values and will differ globally, but a 'green' world will need significant outside-the-box innovation to create the required step-change in the size of the human footprint

3. Sustainable innovation needs to be assessed both with internal criteria such as strategy and profitability as well as with external ones such as ecology and society

4. Dialogue with society is an integral part of sustainable innovation and sustainable development needs to be embedded in the business objectives

5. Sustainable innovation creates value to the company in the form of societal capital; societal capital is part of the intellectual capital and represents the value of the reputation in society and the licence to operate.

– 7 –
Innovation and the CEO

THE FINAL LAW

The future of the company is a key concern for any CEO and innovation is one of the key levers available for responding to the challenges of the future. Particularly at times when there are big changes in the business environment, for instance as a result of globalisation with new competitive positions, emerging clusters of disruptive technologies with new possibilities or changing societal values, with new customer expectations, it is essential that management makes a careful assessment on the strategic use of innovation. Game-changing innovation is a strategic tool that the CEO can choose to use in order to improve the growth potential or the resilience of the company by shifting the business portfolio.

At the start of a new S-curve in the business environment, innovation is the best option for delivering benefits from the emerging new opportunities. What cost reduction can do for the bottom-line in the short term, innovation can do for top-line in the long term. It is up to the CEO to assess how much change in the business or resilience to external changes the company needs in view of its competitive position and the potential changes in the business environment, and which contribution has to come from innovation. The CEO has to articulate the vision on the endgame that has to be played and steer the game-changing innovation effort and the company in that direction.

In small start-up companies, radical innovation is the single business issue on which survival hinges and top management attention is ensured. In large companies, there can be a disparity between words and action. Innovation may rank high in the company's ambitions, but can feature low in the actual business strategy and allocation of resources. Big companies tend to be good in incremental innovation and optimisation to maximise the benefits of economy of scale and global reach, but there are many hurdles to game-changing innovation that are inherent to big companies. It creates competition to existing business interests and it challenges established positions. This conflict has to be managed in a coherent innovation strategy. Game-changing innovation, by definition, will change or stretch the traditional business boundaries, and therefore it has to be managed at corporate level and needs its own, specific governance system. The CEO is the person who can manage this process with the least bias and the best sense of purpose. In the extreme case, when the future of the whole company may be at stake, the CEO has no choice but to be in charge, a truth engrained in all the entrepreneurial start-up companies, but not standard in large companies. The CEO has to champion the innovation process. In the words of the twelfth law of innovation:

Law XII   There is no successful innovative company without a committed CEO

The twelfth law is not based on statistical evidence, and by its nature never will be. There is circumstantial evidence, but the main basis of the law is the straightforward logic that innovation needs direction to create valuable results. And the CEO has to commit the company to that direction. If not, success becomes a matter of random chance, comparable to the chances of a ship arriving at its destination without a captain.

But besides doing the right thing, it is also important to do the things the right way. Besides commitment, the CEO also needs to understand the nature of the innovation process and the way it should be managed for best performance – that is the purpose of the laws of innovation. Managing innovation according to the laws is managing innovation in line with its intrinsic needs and characteristics. Understanding the basic laws of innovation is not a guarantee of success, but it does improve the probability.

The CEO has three main roles: to steer, to protect and to set the rules of the game.

- The CEO has to select the preferred domains for innovation, agree the potential endgames of the options under development, and decide how and when to use them.
- The CEO, as the champion of the innovation capability, has to safeguard the momentum and the continuity of the overall innovation effort through temporary setbacks in the fortunes of projects or of the company; inopportune reduction of the innovation capability destroys intellectual capital and thus shareholder value.

Innovation and the CEO

- Innovation has its own rules and the CEO has to ensure that innovation has its proper space and infrastructure. This can include, for example, stimulating diversity, allowing divergent opinions, supporting conflicting projects, rewarding risk taking and ensuring that the laws of innovation are recognised in the way the innovation process is managed.

The CEO has to perform his roles in a balanced way, combining clear leadership with keeping sufficient distance from the innovation process so as not to influence the internal checks and balances that every innovation project needs to be exposed to during its development period. Most importantly, he should understand that, even with the best effort and judgement, the ultimate winner in game-changing innovation remains unpredictable. The only game the CEO can play is improving the odds and doing it better than the competition.

THE LAWS OF INNOVATION

The twelve laws of innovation cover twelve important aspects of innovation; they state that innovation:

- is a business process
- requires staging
- is opportunity driven
- can be inside- or outside-the-box
- requires external partners
- needs diversity
- is risky
- requires entrepreneurs
- is done to creates options
- creates change
- needs balanced value drivers to be sustainable
- needs commitment from the top

The intent of and the interaction between the laws are illustrated in Table 7.1. The left-hand side of the table below shows how innovation works and on the right-hand side of the table are the associated laws of innovation to provide context and cohesion between the laws and deepen the understanding.

Insight in Innovation

Table 7.1 Understanding the Laws of Innovation

The rationale for innovation	The Laws of Innovation
Innovation is one of the strategic tools that a company can use to shape its future and create a more competitive and resilient position for itself. This requires both a clear strategic direction as well as commitment from the top. The CEO has to understand why and which changes are needed and decide the strategy that should be adopted to achieve them.	*There is no successful innovative company without a **committed CEO*** *The purpose of innovation is to create **desired**, valuable **change***
If the innovation strategy focuses on getting more out of existing assets, the CEO can delegate most of the implementation to the business units; if the strategy aims at rejuvenating or stretching the businesses or radical new ways of doing business, the CEO needs to lead.	*Innovation management distinguishes only **two types of innovation**: inside-the-box and outside-the-box, based on whether or not the pathway to the customer is known at the start*
Innovation and acquisition are the key strategies for creating long-term change and value. Both strategies create new business options, but they differ in the way risk is managed. Outside-the-box acquisition is an abrupt buy-in strategy of intellectual capital to get access to innovation developed by another company, and risk is managed by delaying the time of investment to the time that the customer response can be assessed. Innovation uses internal capabilities and generates intellectual capital, and manages risks by creating a portfolio of options.	*The value of innovation is in the creation of new **options** for the business **and** new **choices** for the customer* *Managing **risks** and learning from failures are the keys to success*

(Continued)

Table 7.1 Continued

The rationale for innovation	The Laws of Innovation
The purpose of the innovation business process is to develop an idea to such a level of understanding that the risks in bringing it to market and making investments are reduced to an acceptable level. This is not easy and the process fails more often than that it is successful.	*Innovation is the **business process** for creating new and insightful ideas and bringing them successfully to the market*
Innovation is an unusual business process because it is the only business process with built-in 'chaos and divergence'. But it is chaos by design to extract the creative powers in the company and give diversity the opportunity to experiment. Chaos needs to be contained and the initial divergent processes have to be changed into a convergent, tightly managed process before the investments can be made. To do this the innovation process is staged and ensures that the capabilities are used in the right place.	***Diversity** is essential to stimulate creativity, creativity needs time and analyses to mature into insight, innovative ideas are based on insight* *The innovation process has three distinct **stages**: ideation, development and investment; the prime requirement for stage 1 is insight, for stage 2 a champion and for stage 3 an entrepreneur*
Any strategy for creating profitable change has to start with customer needs. Innovation creates new opportunities for the business by matching customer needs with existing or newly created capabilities.	*Innovation is **opportunity-driven**; an opportunity is a new combination between (potential) customer needs and (emerging) business and technological capabilities*
In today's world it is not sufficient to justify business innovation on economic grounds only. Sound business strategies need to include ecological impact and societal acceptability. Sustainable development is a business issue and involves the needs of all stakeholders, both in the short as well as the long term. Sustainable innovation needs an equitable business model based on extracting value from the market while simultaneously nurturing it.	*Sustainable innovation has three **value drivers** – technology, business and society – that need to be in balance to create new choices for the customers of today without compromising the options for the future*

(*Continued*)

Table 7.1 Continued

The rationale for innovation	The Laws of Innovation
It is unusual for breakthrough innovation to be developed fully in-house. Developing alliances and networks to create access to the required capabilities is an important element of outside the box innovation. An external orientation is also key to ensure that the interests of all stakeholders are identified. Open windows to the world are a necessary ingredient of sustainable innovation and a channel to bring in diversity.	*Outside-the-box innovation requires* **external partners** *with complementary capabilities to find and develop the best way to the customer*
Only few employees have the bravery and competencies to bring an outside-the-box innovation to the market and are willing to take the risk when the odds are against. Such people are a key asset to the company.	**Entrepreneurship** *is the scarcest resource in outside-the-box innovation*
Key assets like interpreneurs and core innovation capabilities need to be safeguarded to ensure continued access to innovation by the company; innovation needs continuity in effort and commitment by the top.	*There is no successful innovative company without a* **committed** **CEO**

In strategic, game-changing, outside-the-box innovation the first and the last words are for the CEO.

References

1. J. Goudsblom, 1992, *Fire and Civilization*, The Penguin Press
2. W.D. Jr. Phillips and C.R. Phillips, 1992, *The Worlds of Christopher Columbus*, Cambridge University Press
3. R.W. Clark, 1977, *Edison, the Man who Made the Future*, Macdonald and Jane's
4. K. Ellis, 1974, *Thomas Edison, Genius of Electricity*, Priority Press Ltd
5. J.A. Schumpeter, 1934, *The Theory of Economic Development*, Harvard University Press
6. P.F. Drucker, 1985, *Innovation and Entrepreneurship, Practice and Principles*, Heineman
7. A.B. Lovins, 2002, *Small is Profitable*, Rocky Mountain Institute
8. J.V. Buckley, 1998, *Going for Growth, Realizing the Value of Technology*, McGraw-Hill
9. P.A. Roussel, K.N. Saad and T.J. Erickson, 1991, *Third Generation R&D*, Harvard Business School press
10. J. Micklethwait and A. Wooldridge, 2003, *The Company: a Short History of a Revolutionary Idea*, Modern Library
11. G.H. Gaynor, 2002, *Innovation by Design*, Amacom
12. J. Tidd, J. Bessant, K. Pavitt, 1997, *Managing Innovation: Integrating Technological, Market and Organisational Change*, John Wiley and Sons
13. *New Insights and Best Practices in Innovation*, The 1999 Arthur D. Little Ideas Conference
14. H.I. Ansoff, 1965, *Corporate Strategy – An Analytical Approach to Business Policy for Growth and Expansion*, McGraw-Hill
15. G. Hamel, 2000, *Leading the Revolution*, HBS Press
16. R.G. Cooper, 1993, *Winning at New Products: Accelerating the Process from Idea to Launch*, Reading
17. C. Handy, 1999, *The New Alchemists*, Hutchinson
18. J. van der Veer, 2003, *Innovation: From Vision to Reality*, Education without Borders conference
19. R. Pharson and R. Keyes, 2002, *The Failure-Tolerant Leader*, Harvard Business Review
20. G. Hamel, 1999, *Bringing Silicon Valley Inside*, Harvard Business Review
21. W. Bridges, 1995, *Managing Transitions*, Nicholas Brealey Publishing
22. *Lessons from Shell Global Solutions' transformation into a commercial knowledge company*, 2000, Shell International

23. F. Trompenaars and C. Hampden Turner, 2000, *21 Leaders for the 21st Century, How Innovative Leaders Manage in the Digital Age*, McGraw-Hill
24. R.T. Pascale, M. Milleman and L. Gioja, 2000, *Surfing the Edge of Chaos, The Laws of Nature and the New Laws of Business*, Texere
25. A. Morrison, 2003, *Going Beyond the Idea – Delivering Successful Corporate Innovation*, Grist
26. I. Azides, 1992, *Mastering Change*, Azides Institute Publications
27. E.B. Roberts and A.L. Frohman, 1997, *Strategies for Improving Research Utilization*, Technology Review
28. J.B. Quinn, 1995, *Managing Innovation: Controlled Chaos*, Harvard Business Review
29. A.H. van de Ven, D.E. Polley, R. Garud and S. Venkataraman, 1999, *The Innovation Journey*, Oxford University Press
30. A.J. Berkhout, 2000, *The Dynamic Role of Knowledge in Innovation*, Delft University Press
31. G. Altshuller, 1998, *40 Principles: TRIZ Keys to Technical Innovation*, Technical Innovation Center, Inc
32. J. Birkinshaw, R. van Basten Batenburg and G. Murray, 2002, *Corporate Venturing*, London Business School
33. J.E. van Aken and M.C.D.P. Weggeman, 1998, *Management of Innovation Networks as Learning Alliances: Overcoming the Daphne-dilemma*, R&D-Management
34. J.G. Wissema and L. Euser, 1988, *Samenwerking bij Technologische Vernieuwing*, Kluwer Bedrijfswetenschappen
35. J.M. Bardwick, 1991, *Danger in the Comfort Zone*, Amacom
36. D.R. Denison, 1989, *Organizational Culture and Organizational Effectiveness: A Theory and Some Empirical Evidence*, Proceedings of the Academy of Management
37. J.P. Kotter, W. McKnight and J.L. Heskett, 1992, *Corporate Culture and Performance*, The Free Press
38. C.M. Christensen, 1997, *The Innovator's Dilemma*, Harvard Business Press
39. D. Yergin, 1991, *The Prize*, Touchstone
40. S. Howarth, 1997, *A Century in Oil*, Weidenfeld and Nicolson
41. K.J. Leslie and M.P. Michaels, 1997, *The real power of real options*, The McKinsey Quarterly
42. A.K. Dixit and R.S. Pindyck, 2003, *Investments under Uncertainty*, Princeton University Press
43. C.M. Christensen and M.E. Raynor, 2003, *The Innovator's Solution: Creating and Sustaining Successful Growth*, HBS Press
44. G. McConnachie, 1998, *Value Management: The Role of Intellectual Assets and Capital*, The 7th Conference on Global Business Issues of Technology and R&D, MCE-Brussels
45. K.E. Sveiby, 1997, *The New Organisational Wealth, Managing and Measuring Knowledge-Based Assets*, Berrett Koehler
46. L. Edvinsson and M.S. Malone, 1997, *Intellectual Capital – Realizing your Company's True Value by Finding its Hidden Brain Power*, Harper Collins
47. R.S. Kaplan and D.P. Norton, 1996, *The Balanced Score Card*, HBS Press

References

48. J. Mouritsen, H.T. Larsen, P.N. Bukh and M.R. Johansen, *Reading an intellectual capital statement*, Journal of Intellectual Capital, 2001
49. D.J. Skyrme, 1998, *Measuring the Value of Knowledge: Metrics for the Knowledge Based Business*, Business Intelligence Ltd
50. S. Koury, J. Danielle and P. Germeraad, 2001, *Selection and Application of Intellectual Property Valuation Methods in Portfolio Management and Value Extraction*, Les Nouvelles
51. K.G. Rivette and D. Kline, 1999, *Rembrandts in the Attic*, HBS Press
52. J.L. Davis and S.S. Harrison, 2001, *Edison in the Boardroom*, John Wiley and Sons
53. E. Tenner, 1996, *Why Things Fight Back: Technology and the Revenge of Unintended Consequences*, Knopf
54. P.E. Gray, 1989, *Technology and Environment*, National Academy Press.
55. R.L. Carson, 1962, *Silent Spring*, Houghton Mifflin
56. S.L. Hart, 1997, *Beyond Greening: Strategies for a Sustainable World*, Harvard Business Review
57. P. Hawken, A. Lovins and L.H. Lovins, 2002, *Natural Capitalism, Creating the Next Industrial Revolution*, Back Bay Books
58. W. McDonough and M. Braumgart, 2002, *Cradle to Cradle: Remaking the Way we Make Things*, North Point Press
59. L.R. Brown, 2001, *Eco-Economy: Building an Economy for the Earth*, W.W. Norton and Company
60. S.L. Hart and M.B. Milstein, 1999, *Global Sustainability and the Creative Destruction of Industries*, Sloan Management Review
61. A. Grübler, 1998, *Technology and Global Change*, Cambridge University Press
62. M. Wackernagel and W. Rees, 1995, *Our Ecological Footprint: Reducing Human Impact on the Earth*, New Society Pub
63. J. van der Veer, 1998, *The Greenhouse Challenge — Dialogues, Decisions and Delivery*, Conference on "Greenhouse Gas Control Technologies"
64. A. Toffler, 1989, *The Third Wave*, Bantam
65. OECD secretariat paper, 1998, *Foresight Methodologies and the Environment*
66. A. Dearing, 1999, *Have we the foresight for sustainable development?* Foresight
67. J.J. van Duijn, 1982, *The Long Wave in Economic Life*, Unwin Hymen
68. P. Schwartz, P. Leyden and J. Hyatt, 2000, *The Long Boom: A Future History of the World*, Orion
69. W. Buckland, A. Hatcher and J. Birkinshaw, 2003, *Inventuring – Why Big Companies Must Think Small*, McGraw-Hill

Index

A

adaptive innovation, 24, 41
adding value, 6
ADL, 22, 110
alliances, 31, 50, 54
America, 2
Amerigo Vespucci, 3
Ansoff, 23

B

backcasting, 130
benchmarking, 41
bottom-line, 40, 102, 137
bottom-up, 29, 30, 43, 63
breakthrough, 22
bridge-building model, 17, 20, 124
bridge-building innovation model, 15
Brundtland, 117
business model, 38, 40
business plan, 2, 8
business process, 7, 45, 70, 139, 141

C

capital, 140
carrying capacity, 117, 121
cascading model, 20
champion, 8, 10, 18, 27, 37, 71, 138
change, 118, 122, 139
chaos, 54, 141
classes of innovation, 21
classic cascading model, 17
classic innovation model, 11, 12
classification, 22, 29, 77, 78
Columbia, 3
Columbus, 2, 3, 8–11

comfort zone, 29, 34
committed CEO, 138, 140, 142
company strategy, 27
competitive innovation, 24
Cooper, 28, 48
corporate venture units, 77, 78
cost-benefit, 15
creativity, 1, 5, 6, 10, 50, 51
customer, 1, 38, 91, 97, 116, 120, 125, 134, 140
customer capital (external capital), 108, 112, 135
customer domain, 92
customer value propositions, 2, 7, 35, 67

D

Darwin, 135
diffusion, 16
discovery, 2, 3, 7, 16
discoveries, 11
disruptive innovations, 79
diversity, 51, 70, 134, 139, 141
Dow Chemical, 110
drive, 89
Drucker, 6, 50
dynamics, 41, 44

E

ecosystems, 117, 121, 122
Edison, 4, 8, 9, 10, 11, 116
Edison Electric Illuminating Company, 4
Edison Electric Light Company, 4
efficient frontier, 104
emotions, 87
entrepreneur, 4, 5, 10, 54, 139

INDEX

entrepreneurship, 5, 6, 69, 73, 74, 89, 90, 142
equity, 117, 120
external innovation, 35
external stakeholders capital, 135

F

fair market value, 110
fast results delivery, 30
feasibility, 7
feedback loop, 14
financial capital, 107, 112, 135
fire, 1

G

Gary Hamel, 24
GameChanger, 30
game-changing, 142
game-changing innovation, 25, 27, 28, 58, 66, 88, 138
grassroots, 50
grey space, 62

H

human capital, 107, 108, 112, 135
human footprint, 121, 122, 136
Humphrey Davy, 4
hydrogen, 126

I

idea, 6, 7, 10, 46, 51
impact zone, 29, 34
incremental innovation, 24, 25, 27
infrastructure, 5, 85, 86, 89
innovare, 5
innovation, 4–5, 69, 89
innovation capabilities, 108, 142
innovation capability, 106, 116, 138
innovation culture, 68, 69, 71, 89
innovation domain, 92
innovation environment, 17
innovation eras, 18
innovation funnel, 47
innovation infrastructure, 65, 66, 69, 71
innovation management, 2, 11, 15, 69
innovation manager, 58, 65, 71
innovation models, 17
innovation options, 73
innovation platforms, 64
innovation portfolio, 63, 71, 105, 116
innovation process, 1–3, 10, 11, 17
innovation spectrum, 21
innovation strategy, 61, 71
innovation supply chain, 12, 14, 17, 18, 20, 29, 46, 59
innovation zones, 33
innovator, 5, 8
innovator's dilemma, 102
inside-the-box, 23, 24, 26, 27, 36, 43, 55, 67, 73, 99, 139, 140
insight, 6, 10, 51
insightful, 6, 89
inspiration, 1, 90
intangible assets, 106, 109, 111
intellectual assets, 107, 108, 112, 135
intellectual capital, 106–108, 110, 112, 116, 135, 138
intellectual property, 57, 81, 108, 112, 116, 135
internal innovation, 35
internal organisational capital, 135
interpreneur, 76, 90, 142
invention, 3, 5, 7, 11, 69, 89
inventor, 4, 8, 18

J

joint venture, 54, 75, 78

INDEX

K

King Ferdinand, 2, 3
King Johan II, 3
Kondratieff, 130

L

law, 6, 10, 16, 26, 33, 51, 70, 74, 91, 118, 125, 137, 138
laws of innovation, 1, 6, 140
leakage, 32, 43
leverage, 43
lifecycle, 19, 120, 130

M

management techniques, 6, 12
managing innovation, 1, 45
Marcus Samuel, 94, 95, 97, 100
market value, 108
momentum, 42, 59
Murex, 94, 95, 97, 98

N

Neue Kombinationen, 6
new combinations, 6, 45, 97
new-to-the-market, 22
NPV (net present value), 99, 100

O

opportunity-driven, 16, 20, 139, 141
options, 73, 74, 81, 91, 139, 140
option matrix, 113
option value, 101, 116
organisational assets, 107, 108, 112, 135
organisational capital (internal capital), 107, 112

outside-the-box, 23, 24, 26, 27, 33, 36, 37, 43, 55, 66, 67, 73, 88, 99, 116, 136, 140, 142

P

partners, 14, 33, 139, 142
perspiration, 1
platform, 85
portfolio, 63
portfolio of options, 102, 140
process of innovation, 11
product, 22
product strategy, 27
prototype, 7
Ptolemaeus, 3

Q

Queen Isabelle, 3

R

R&D management, 15
Rembrandt, 116
research management, 66
residual risks, 71
revolutionary innovation, 24
reward, 68, 104
risk, 1, 2, 8, 26, 70, 71, 78, 81, 97, 104, 139, 140
Rockefeller, 97

S

S-curve, 19, 85, 102, 137
scenarios, 51, 130
Schumpeter, 5, 15, 19
scientific curiosity, 18
serendipity, 60
Shell, 6, 37, 51, 78, 97, 126, 127, 130

INDEX

Shell Global Solutions, 38, 40, 42, 112
Shell Hydrogen, 78, 79
Silent Spring, 120
Silicon Valley, 8, 133
societal capital, 134–136
stages, 7, 8, 37, 40, 49, 141
stage 1, 46, 49
stage 2, 52
stage 3, 54
stage-gate, 30, 49, 70
stage-gated process, 48, 139
stakeholders, 50, 62, 89, 117, 120, 142
standards, 5, 85, 86
Standard Oil, 95
stop-go, 55
strategic alliances, 14
strategy domain, 92
structural capital, 107, 108, 135
supply chain, 61
supply chain management, 71
supply chain process, 7, 70
sustainable development, 117, 122, 141
sustainable innovation, 117, 119, 120, 124, 125, 131–132, 136, 141

T

3M, 84
technology development, 67
technology enabled innovation, 11
technology factor, 110
technology management, 66
technology road mapping, 131
time to market, 14, 82, 97
tollgates, 48, 49, 54, 56, 57, 60, 83
tollgate I, 56, 57
tollgate II, 56, 57
tollgate III, 56, 57
top line, 102
top-down, 29, 30, 43, 63
Toscanelli, 3
transformation, 35, 40, 44
trial-and-error, 18, 20
two types of innovation, 140
types of innovation, 27

U

United Nations, 117

V

valley of death, 37, 73, 74
value, 90, 91, 106, 110, 112
value domains, 92, 94
value drivers, 97, 125, 139, 141
value matrix, 115
value of an option, 99
value of innovation, 98, 116
value/cost ratios, 111
value-driven model, 124
venture capitalist, 54
venture unit, 81, 106
vision, 10, 39, 40, 87, 89

W

Waldseemüller, 3
WBCSD, 117
white space, 23, 61, 62
Wright brothers, 11